CITIES
IN THE
21st CENTURY

D0325283

Volume 23, URBAN AFFAIRS ANNUAL REVIEWS

CITIES
IN THE
21st CENTURY

Edited by
GARY GAPPERT
and
RICHARD V. KNIGHT

Volume 23, URBAN AFFAIRS ANNUAL REVIEWS

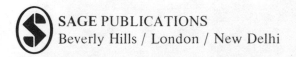

SAGE PUBLICATIONS
Beverly Hills / London / New Delhi

Copyright © 1982 by Sage Publications, Inc.

For information address:

SAGE Publications, Inc.
275 South Beverly Drive
Beverly Hills, California 90212

SAGE Publications India Pvt. Ltd.      SAGE Publications Ltd
C-236 Defence Colony   28 Banner Street
New Delhi 110 024, India   London EC1Y 8QE, England

Printed in the United States of America

Library of Congress Cataloging in Publication Data

Main entry under title:

Cities in the 21st century.

   (Urban affairs annual reviews ; v. 23)
   Bibliography: p.
   1. Cities and towns—United States—Forecasting—
Addresses, essays, lectures.   2. Twenty-first century—
Forecasts—Addresses, essays, lectures.   I. Gappert, Gary.
II. Knight, Richard V.   III. Series
HT108.U7   vol. 23   307.7'6s   [307.7'6'0973]   82-16858
[HT123]
ISBN 0-8039-1910-7
ISBN 0-8039-1911-5 (pbk.)

FIRST PRINTING

Contents

This volume is dedicated to James Rouse and Henry Reuss, both of whom have contributed a great deal to the renaissance of American cities and the advocacy of an urbane style of living.

---

*Because I know that time is always time
and place is always and only place
and what is actual is actual only for one time
and only for one place
I rejoice that things are as they are . . .*

T.S. Eliot, *Ash Wednesday*

1

Future Urban America:
Post-Affluent or Advanced Industrial Society?

GARY GAPPERT

☐ THERE ARE SOME OBSERVERS who believe that the future, especially the American future, is out of control. No one is in charge, and although there is no conspiracy, even the emerging economic anarchy seems poorly organized (Schwartz et al., 1982). Others feel that the future is technologically determined and inevitable. Toffler (1981) proposes a society of electronic cottages, and Ferguson (1980) reveals the aquarian conspiracy. To these observers the apparent social disarray is just the prelude to a new stage of higher consciousness and cultural achievement.

More seasoned observers feel that the future is a rich and virtually indigestible stew. Gray (1982) has referred to a technological barnyard. Marien (1976) has tracked over sixty definitions or descriptions of the projected society of the future. They range from "post-capitalist" to "postwarfare" and include such images as the "one-dimensional society" and "beyond the stable state." These kinds of images or metaphors represent attempts to bring some kind of conceptual unity to the myriad bits of the future emerging around us. Lifton (1976) has suggested that as the sense of change intensifies, the boundaries between history and further evolution are obscured. People seem to be plunging ahead into an unknowable process

AUTHOR'S NOTE: *The ideas in this section have been developed over a number of years and reflect the thoughts of numerous colleagues. A forty-minute slide presentation of this material was screened at the annual meeting of the Urban Affairs Association at Philadelphia in April 1982 and is available for other forums.*

without a clear destination. Then they "suddenly discover, swirling about, the total array of images created over the full course of their historical and evolutionary past." He concludes:

These images become an elusive form of psychic nutriment, to be ingested, metabolized, excreted and, above all, built upon and recombined in some kind of vital process [Lifton,1975: 137]

In developing a perspective on the future of cities in our society, we need to begin to play with a number of images in order to achieve a similar symbolic reconstruction or theoretical integration with respect to the role of urban development in a changing society. Will society be different in the year 2000?

POST-SOMETHING BUT POST-WHAT

For the last twenty years or so Daniel Bell has befuddled us with his notion of a postindustrial society (1973). This metaphor has become an extremely pervasive framework for analyzing the conditions and problems of contemporary society. The post-something form of analysis is discussed by Lifton as a necessary condition for social or cultural revitalization:

Consider for instance, the widespread inclination to name and interpret the contemporary man or woman not in terms of what he or she might actually be but rather in terms of what has been—that is, in terms of what we have survived. We speak of ourselves as post-modern, post-industrial, post-historic, post-identity, post-economic, post-materialistic, post-technocratic, and so forth. There are pitfalls in this way of managing the present (or the future) after what no longer is (or will be), but the terms have an authentic source in the sense of survivorship, present or anticipated, that so pervades our deepest image of ourselves [1975: 135].

Lifton suggests that there are several "kinds of technological and cultural holocausts, real and anticipated" hovering over contemporary society. In his diagnosis a sense of both loss and survival creates an urgency for innovation and a "task of resymbolization" as "something close to a mass experience." Images of "economic holocaust" became common in the late 1970s and still persist today, notwithstanding the conservative euphoria over supply-side economic initiative.

The framework of a post-affluent society developed in the mid-1970s (Gappert, 1979) was one attempt to create a model to assess the myriad trends and events associated with the fundamental dislocation of the American economy caused by OPEC seizing the international economic order.

The assumption of the post-affluent framework is that the economic events of 1971-1974 represent a significant turning point and developmental shift in our society. These events—price controls, the oil boycott, the dramatic increase in energy costs—triggered the start of a post-affluent transition. This transition was preceded by the beginnings of a postindustrial shift out of traditional manufacturing employment. The significant beginning of this postindustrial shift can be symbolized by the Soviet launching of Sputnik in 1957. The American response to that event led to the National Defense Education Act, and to the acceleration of space and defense programs out of which both new technologies and new sunbelt cities have developed.

The postindustrial framework, however, has lost conceptual vigor as the American economy has evolved toward a further stage of industrial development. The usefullness of the post-affluent framework is also likely to decline if a reconstruction of the American economy is achieved. But we are still in the midst of a transitional phase. The processes that Schumpeter referred to as "creative destruction" are still rampant. A restructured economy is not yet here.

THE POST-AFFLUENT TRANSITION

It is likely that the seven elements of a post-affluent framework will remain relevant for the rest of the 1980s. These elements are:

(1) The almost overwhelming significance of the baby-boom population—the bulge generation.

(2) The emergence and recognition of a post-affluent consciousness.

(3) The recognition, if not the acceptance of the transcendental nature of many wants and needs.

(4) Innovative tinkering and possible reforms in households and work places, including social-economic counseling and job redesign.

(5) The gradual development of synergistic lifestyles in an evolving androgynous culture.

(6) Discontent and disruptions with respect to agreement on a just distribution of income among different socioeconomic groups.

(7) Patterns of uncertainty and indecision among large-scale organizations, including the federal government.

The so-called baby-boom population—the bulge generation—gave us the onslaught of 17 year-olds in the late 1960s and an onslaught of new workers in the 1970s. This generation has now started its own baby boom (New York *Times*, September 2, 1980) as its women race against the clock to have children. This young adult population occupies the large apartment complexes in metropolitan areas and will represent a sizable demand for urban housing in the decade ahead.

Two other demographic issues are also worth noting. For instance, the baby-boom population is preceded by the "depression-dent" generation. The population cohort born between 1934 and 1935 is rather thinly populated and may not be able to provide sufficient high quality senior managers as America prepares for the 1990s.

On the other hand, the second half of the baby boom population is still in school or is just entering the labor market. This has created a prolongation of adolescence as young adults born between 1955 and 1961 scramble for positions in the adult world. (For example, when one of the Kennedy children married sportscaster Frank Gifford's daughter in New York's St. Patrick Cathedral, the commentary ended: "Both youngsters are 24.") The movement of "adult youngsters" metropolitan areas is a new market phenomena that is likely to have unforeseen consequences, especially in lifestyle arrangements, marketplace behavior, and value development.

Another demographic reality is the growing number of elderly people. The age-65-and-over cohort group will grow from about 25 million in 1980 to 30 million in 1990, but only to about 32 million by the year 2000 (the depression-dent population). But the ratio of the elderly to the rest of the population will remain stable at about 11 to 12% for the rest of this century. A sharp increase in the proportion of the elderly population will not occur until the 2010s when the baby boomers reach retirement age.

The second element of our post-affluent society is simply the growing awareness of post-affluent consciousness. Yankelovich (1981) points out that

> 72 percent of the public now believes that "this land of plenty is becoming a land of want." ... A majority of Americans now state that ten years from now they don't think they will be able to buy a new car or to own a home of their own.

He continues:

> Even the 40 percent or so of the public whose incomes have more than
> kept up with inflation and are therefore objectively better off than they
> were, don't feel better off.

Affluence (from Latin *affluens,* present participle of *affluere,* to
flow to), if it means anything at all, means some control over
discretionary income. It means having a regular surplus at the end of
each month and at the end of the fiscal year. As I wrote in *Post Affluent America:*

> In 1973-75 the expectations of the American middle class were shat-
> tered. Within those two years, the purchasing power of the dollar
> shrank by nearly 20 percent, the value of household savings and other
> wealth and assets declined about 12 percent, and the size of household
> debts rose by over 18 percent [Gappert, 1979:46].

Since that time both our governmental and personal indebtedness
have increased, as have the number of our Eurodollars and
petrodollars held abroad. John Nevin, president of Firestone Tire and
Rubber Company, has recently reminded us that

> Huge trade deficits are a recent phenomena in American history.
> During the five year period immediately preceding the 1973 oil crisis,
> American trade deficits averaged only $1.1 billion annually. During the
> five year 1977-81 period, American deficits averaged $27.9 billion
> annually [1982].

The involuntary decreases in the standard of living have partially
been absorbed by the transition of many families from one-wage to
two-wage households. But, unless we change the child-labor laws, we
have run out of potential new wage earners in most families, and
multiple job holding has become another consequence.

Unfortunately, the significance of a post-affluent consciousness is
not widely understood by our national leadership. The current version
of supply-side economics is still oriented to the assumption that a shift
of resources from the public sector to the private sector will guarantee
a substantial increase in our standard of living. A real supply-side

economics would emphasize the need to reduce and postpone consumption, to increase savings and investment, and to reduce our trade deficit. Otherwise, everything points to a further decline in living standards.

This brings us to the third element of a post-affluent society, the changing and evolving nature of economic needs and wants. The American consumer, faced with the shrinkage of his or her discretionary dollar, is engaged in an agonizing reevaluation of basic needs and wants. The downsizing of the American automobile is one sign of this, as is the dramatic decrease in oil imports. In 1977 the United States imported 8.5 million barrels a day; by 1981 oil imports had dropped to 6.1 million barrels.

The rest of the 1980s will see the comparable downsizing of the American home. Between 1960 and 1980 residential house-heating bills comprised 1.5% or less of real disposable income in American households. Total decontrol in 1985 will increase the house-heating portion of disposable income to about 3.2%. The impact will be even greater at the lower-income end of the scale where households earning less than $15,000 a year will spend 7 to 14% disposable income on heat.

The downsizing of the American house has already implicitly begun as members of the baby-boom population remain tied to the large apartment complexes ringing our central cities. The growth of personal storage centers in adjacent areas is a good example of the marketplace responding to opportunity.

Bruce Stokes (1981) of the Worldwatch Institute has analyzed housing prospects, including an increase in higher density urban living and the growth of a market for "fourplexes" and "sixplexes"—large houses divided into four or six separate units. As is already common in Europe, many units will be sold with several unfurnished rooms, allowing the purchaser to finish them when time and resources permit.

Other reevaluations are also going on in the marketplace as distinctions are made between absolute needs and those that are relative (Hirsch, 1977). A secondary consequence of these involuntary market adjustments are the growth of conspicuous frugality and voluntary austerity as social values.

Elsewhere, we have contrasted a Hobbesian future with a more Emersonian future. The Hobbesian future will be "nasty, brutish, and unstable" (Blair and Gapper, 1976). The Emersonian future is characterized by a view of material wealth as a means for attaining more satisfactory lifestyles and varieties of human experience rather than as an end in itself. Among the features of the Emersonian future are (1) a flourishing human potential movement; (2) a variety of alternatives to traditional work; (3) intolerance of conspicuous consumption and waste; and (4) a broadening of the educational process.

The fourth element in our post-affluent framework is the relationship between reforms and readjustments in both households and workplaces. The quality-of-work issue is not new. What is new is that the decline of American productivity has led American business to reexamine some of the job-enrichment literature of the 1950s and 1960s that was ignored when American prosperity seemed boundless. Simultaneously, American labor is also being forced to reexamine some of their demands. Some of the "give back or shut down" issues are tied directly to cost-cutting productivity issues. Others are tied to the realization that the workplace is also a social environment. Still others are related to the motivational environment in which the new knowledge workers (and many factory jobs are now knowledge-intensive) are best utilized.

These quality-of-work issues also exist in the new social context of the two-paycheck family and the single-parent household. As Linden (1971) points out, our household time budgets are even more inflexible than our fiscal budgets. Part of the surprising success of the fast-food industry is attributed to the growth of the two-paycheck family. The emergence of concerns with both stress management and parenting skills can also be attributed to pressures on our time budgets. Part of the attraction of the electronic cottage lies in the potential flexibility of a continuous readjustment of time allocations between different forms of work.

A fifth element of post-affluence is the gradual development of synergistic lifestyles in an evolving androgynous culture. Social and economic roles for both sexes will be more balanced and more evenly distributed. As Young and Willmolt (1973) have written, "one of the greatest social changes of modern times is the move towards symmetry between the sexes." They refer to the "mid-century alliance of family and technology" and note:

> In the course of time the family has re-established a new kind of primacy, not as the unit of production so much as the unit of consumption [p. 14].

The readjustment of family structures and gender roles is of course mirrored by the new structure of the work force. It is projected that by 1988 almost half of the labor force will be women. This is one way to introduce another post-something construct: that the emerging social system can be usefully characterized as "postmasculine," or "postmacho." Society is going to become much more androgynous and will have a more even distribution of the characteristics of both sexes. However, symmetrical families, the androgynous workplace, and the assertion of equal rights are also forces that generate countertrends. Male anxiety and new concerns about the nature of masculinity are emerging and will also be social forces in the decade ahead.

This leads us to the sixth element, which is the continuation of discontent with respect to equity issues. "Who gets what and why" will be a question that emerges in different forms in the 1980s. Stockman's statement that supply-side economics was just old-fashioned trickle-down economics is one manifestation of that. Another emerging issue which will be stronger in the 1980s is the question of regional equity between the snowbelt and the sunbelt. The latter half of the baby-boom population will further fuel the migration to southern and western states.

Another interesting issue is the language of the President's Commission For a National Agenda for the Eighties. Its report refers to "a nearly permanent urban underclass." The word "nearly" hardly qualifies the word "permanent." But the nature of that urban underclass may be more dynamic. What is less clear is whether the growth of substantial structural unemployment is going to transfer into working-class anger or only resignation. Our colleagues in Youngstown refer to the "stunned immobility" of the workers when a plant closes. The full effects of economic dislocation are not yet apparent.

This brings us to the last element of a post-affluent society, which is the pattern of uncertainty and indecision among our major institutions, both public and private.

This element has many manifestations, including the turnover of our political leadership structure in the 1980 elections. But this turnover is paralleled by the now almost weekly dismissal of chief executive officers by American corporations. According to Eugene Jennings, professor of management at the University of Michigan, the dismissals of corporation presidents since the late 1970s is unprecedented in the history of American business.

This leadership gap or apparent leadership failure is probably strongly related to the growth of environmental turbulence. New public expectations as well as shifting values and markets have reduced the predictability for organizations in both the public and private sector. "Business as usual" is a phrase that rarely applies anywhere.

In addition, the 48 to 65 year-old cohort is a generation that has almost been exhausted by change. They grew up in the Depression, they survived World War II, and they saw the "peace and prosperity" under Eisenhower shattered by racial conflict, assassinations, limited warfare, and the social changes of the 1960s. But this is also the group that managed growth in the 1960s and contributed to the invention of futures research, strategic planning, and technology assessment in the early 1970s.

We may repent of some of these constructions of the post-affluent future. The post-affluent portrait is not a photograph, and was begun as a series of drawings in 1972-1973 (Gappert, 1973). The paint is still wet, and this portrait may be redone, the experiences of the future may change the meaning of it for us (pentimento!). However, we must start with some picture of the future.

Another emerging but elusive picture of the future is that of an advanced industrial economy. There is a faint hint that we might be experiencing the start of another long wave of economic development. Could the American economy be on the verge of another great surge of prosperity based upon the new information and biological technologies? The post-affluent prospect might only be a temporary readjustment.

THE ADVANCED INDUSTRIAL ALTERNATIVE

A model of an advanced industrial society could also be developed as an alternative framework for the interpretation and projection of emerging trends and realities in American society. The glass may be half full, not half empty.

As shown in Figure 1.1, there are at least three different ways to perceive the social-economic future. The first perspective, System X, assumes that issues of economic productivity and efficiency will lead to both fiercer competition and greater disappointment unless the vitality of the national economy dramatically improves. In this perspective our cities will continue to be sites of social hostility and underclass turmoil.

The second perspective, System Y, assumes that although the growth rate may decline, the underlying strength of the national economy is undiminished even though individual gains take longer to achieve. In this perspective individual and household pursuit of purely economic objectives are postponed, deflected, or sublimated by ac-

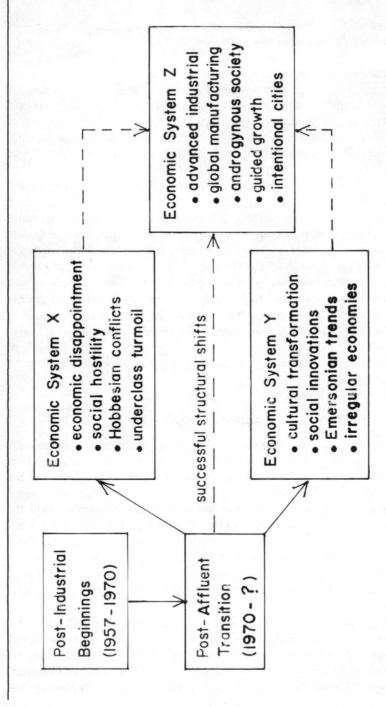

FIGURE 1.1   Perspective on Societal Futures

tivities associated with quality-of-life concerns. Cities and suburbs both contribute to the growth of self-reliance and a self-directing social transformation if this perspective predominates.

Substantial elements of both System X and System Y exist today as part of a constrained economy. The third perspective, that of a System Z, is more elusive and harder to establish. The key assumption of this perspective is that a new kind of technological-industrial progress is possible, and probable. It assumes that a second industrial revolution is at hand based upon technologies that are (1) knowledge intensive, (2) resource conserving, and (3) environmentally benign.

Only a handful of observers have begun to develop a conceptual orientation that contributes to this third perspective that assumes a new techno-social economic structure is forthcoming. Blair and Gappert broached the concept of a transformation in a paper prepared for the Joint Economic Committee (1976). Bradshaw has used California as a leading example of a prototypical advanced industrial society (1979, 1980). Rejecting the term postindustrial because there is no evidence to support the argument of Bell and others that the industrial sector is declining, he has emphasized the following characteristics as advanced-industrial:

—increases in technologically sophisticated and knowledge-intensive industrial employment,
—the growth of the service sector almost entirely at the expense of agricultural rather than manufacturing employment,
—the growth of a massive and varied educational system,
—greater social interdependence.

He claims that at every stage California has been shaped by massive reliance on the most advanced technologies available in the world at that particular time, and has been an innovator, with its products, practices, and lifestyles diffusing to other places in the United States and the world. He argues that California is one of the places at the cutting edge of social change, and as such is a good laboratory in which to examine, more closely than would be possible in the nation as a whole, theories of future social development.

Hamrin (1980) lays out in a comprehensive fashion the elements of an economy centered on information activity. He claims that "the adjective 'information' is more precise than 'service' or 'post-industrial' for describing the fundamental dynamic element that will shape the economy and society of the future." A critical element in Hamrin's analysis is that the ultimate limits to growth in an advanced industrial society "will basically reflect not resource scarcities but a steadily rising preference for noneconomic endeavors and satisfactions." Developing what he calls a "transindustrial world view," Hamrin also proposes a new paradigm for economic growth in which productivity and technological change are important factors along with capital, labor (human resources), and land (natural resources and energy). The guiding norm or ethic within his model of an advanced industrial society is "selective growth" as opposed to traditional forms of simple quantitative measures. This suggests that institutions need to make more holistic decisions in the context of long-range policy frameworks. Although he is pessimistic about the prospects for an immediate adoption of such a new paradigm, his use of a return-on-resources decision-making model for corporations confirms the need to explore different forms of innovative management more appropriate to the changing realities in an advanced industrial society.

Hirschhorn (1982) offers yet another perspective. His reference point is the idea of "advanced capitalism" as distinct from traditional forms. He claims that "there is increasing evidence that capitalism is entering a new historical phase of disaccumulation in which the wage labor system is contracting and the accumulation of variable capital is coming slowly to a halt." Coincident with this, a new work system, new productive resources, and new sources of productivity are emerging that are not consistent with traditional categories of economic classification. The problem that Hirschhorn then goes on to explore is the ways in which older social forms and institutional structures may stifle the developmental processes emerging in an advanced capitalistic or industrial society.

Hirschhorn attempts to develop a model of the advanced industrial labor process based upon seven related hypotheses. A key issue for him is the distribution of intelligence and information in work systems and the ways in which advanced technologies diffuse and change the management functions. He concludes that

> It is urgent that research within this framework be developed so that we do not pursue obsolete lines of thought on the one side and are prepared for genuinely new and emergent possibilities, and conflicts, on the other.

Knight (1979, 1980), analyzing employment expansion in several older metropolitan regions including Cleveland, has also developed a perspective on industrial transition and transformation. Knight's model is based upon the assumption that the large industrial corporations have reached a stage where their "technology transfer" functions have expanded their multinational status in a global economy that has grown by a factor of three in the last decade, while world trade has increased by a factor of seven. This has both generated new streams of real income into the region and the export of traditional production jobs to other states and nations. (Knight's model is further elaborated in Chapter 3.) Other efforts to formulate the social and economic consequences of a new wave of technological development can be found in Ouchi (1981), Mensch (1982), and Masuda (1980). All offer hints of a new systemic possibility.

SYSTEM Z AND CITIES

A System Z perspective that suggests the possibilities and the prospects of a new and higher stage of industrial development might include these seven elements:

—a growth in national productivity due in part to the maturing of the baby-boom labor force (Jones, 1980);

—the futher utilization and expansion of knowledge-intensive and resource-conserving technologies (Masuda, 1981);

—a reorientation of consumer behavior toward a greater appreciation of frugality and self-reliance (Johnson, 1978);

—the acceleration of new forms of workplace organization, with social objectives and compatible recurrent learning requirements;

—the articulation and application of new personality theories that recognize the multidimensional nature of the self in an androgynous society (Ogilvy, 1979; Gappert, 1979);

—the development of new mechanisms of income redistribution, including unorthodox sources of second income (Blair, this volume);

—the formulation of new styles of management and new forms of policy development that recognize emerging global realities (Drucker, 1980).

These elements of an advanced industrial society—of a new kind of social economic system—should be elaborated further. They suggest that growth and development could be directed or guided toward some definite social and economic goals at different levels of our society over some period of time. Unfortunately, such a system of guided growth is not consistent with much of the current national

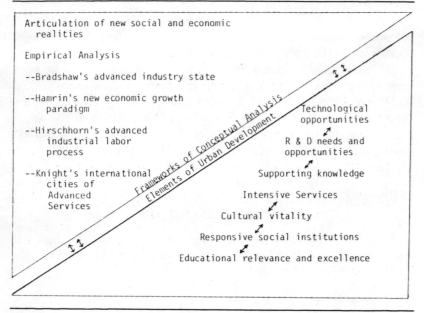

FIGURE 1.2   The Advanced Industrial Economy and its Urban Opportunities

rhetoric. Within such a system individual cities might have different roles and functions that would be explicitly recognized and supported. These would be "intentional cities." In the absence of any policy that advocates, or even just articulates, the nature of such a different socioeconmic system organized around new technological opportunities, the question is: Will such a system emerge anyhow?

One answer is: Let's wait and see. Another is: It's happening already. But in either case even the mere possibility of an advanced industrial economy and society offers a more dramatic set of opportunities to cities than the grim sense of decline associated with the postindustrial construct. Figure 1.2 is a simplistic attempt to suggest that elements of urban development, or areas of opportunity, can be derived from an appreciation of several of the conceptual frameworks embedded in the idea of an advanced industrial society. As someone once said, there is nothing as practical as a good theory.

If one accepts the realities of the current post-affluent condition, but still appreciates the dynamic qualities of American society, its economic strengths, and its technological opportunities, then it becomes possible to concentrate on the identification and elaboration of the role of cities in the new order of things. But many are not yet either cognizant of, or comfortable with, the new realities and opportunities.

According to Phillips:

> Two decades of political and economic trauma have brought this country to a point of considerable risk.... In an era of upheaval like our own, there is no going back, no real way to recapture the past.... To future historians, the early 1980s are almost certain to mark a transition to a new politics, a new economics, and a new philosophy of governance [Phillips, 1982].

But the transition is not yet complete. Therefore, it is still too difficult to forecast the exact nature of urban development. Using the System Z perspective, however, it is possible to formulate a hypothesis that urban development is likely to be led by those cities that can be characterized as knowledge-intensive. Some cities might be characterized as knowledge-adequate, while still others might be knowledge-deficient. There is an advertisement that asserts that the successful businesses of the future will be those that learn to manage information. It may be that the successful cities of the future will be those that learn to manage knowledge and its development, transmittal, and utilization. Cities that embrace a broad strategy for participating in the technological opportunities that appear to lie ahead will also require an emphasis on the renewal of its human capital and on the innovative qualities of its institutions (Niebuhr, 1979).

Years ago Wirth formulated the concepts of urbanism around the notions of size, heterogeneity, and density. A more contemporary conception of urbanism might also include the notion of "institutional vitality." What is missed in some of the more simplistic interpretations of the communications revolution is that the so-called explosion of information just creates more ignorance unless that information is transformed into knowledge that can be utilized by individuals in both systematic and spontaneous ways. That transformation of information into knowledge is a major role for all of our institutions in both the public and private sectors. The research and development office, the marketing division, the urban university, the municipal planning agency, the trade association—all the intermediate institutions and organizations that operate within the urban region— are what help the individual citizen and worker function effectively in a complex society. It is these intermediate institutions that operate at the level of the urban scale that are the single most important interface between the local citizen and the national society and the global economy.

In an advanced industrial society, in the so-called System Z, the superior and successful cities of the late twentieth and early twenty-first century will be those that have developed institutions that can (1) manage knowledge, (2) promote innovation, and (3) guide scarce resources toward strategic priorities. Such cities are likely to have strong external links to significant economic and political networks as well as good interorganizational links at several levels within the local region.

CONCLUSIONS AND SPECULATIONS

John Dewey once wrote, "In any experiment of thinking, premises emerge only as conclusions become manifest." It is difficult to think about the manifestations of cities in the twenty-first century. Although many of us have 25 or 30 year mortgages, we seldom have reason, or the inclination, to think in that long a time framework. But the next 25 years will take American cities from 1982 to the year 2007 A.D. What does that time include? It probably includes:

—Six more presidential elections;

—A period of time equivalent to that from the launching of Sputnik (1957) to the present;

—The aging of the baby-boom generation through their midlife years;

—The development of an increasingly global economy;

—A generational transfer of power.

It is also a period of time in which a new kind of city will develop, a world-class city, a city that will be transnational in many of its transactions.

The manifestations of urban development during that time are not impossible to define. For the next decade or so, a young (under 40) adult learning force will invade our cities, with growing demands for continuing education and training. A new national economic expansion driven by information-intensive technologies and the expansion of global business services should be underway by the early 1990s. The marketplace reorientation to conservation, self-reliance, amenities, and quality-of-life issues will be further developed.

Cities will become more polynucleated, with the development of more multiple-activity centers, multiple-use megastructures, and medium-density planned housing developments. Large tracts of substandard housing are likely to be replaced by new housing, while older, solid housing stock will continue to experience gentrification. Urban farming and woodlots will not be unusual.

Individual housing units will find their design influenced by the conversion of family rooms into work-study centers and by the use of passive solar techniques. The distinction between work and leisure will begin to break down for several kinds of workers. At the same time the growth of involuntary leisure time among the elderly, the young, and the semiskilled will require new social inventions for its constructive utilization. The more efficient and effective use of urban space, both internal and external, private and public, will follow some of the innovative design arrangements in European cities. The postmodernist, neo-Romantic schools of urban architecture are likely to flourish as several generations of urban citizens become more urbane consumers.

Any elaboration of patterns of urban development will require that particular cities must design their own specific scenarios for taking advantage of emerging opportunities within the real constraints present. Urban speculations can only become urban conclusions if there is a community consensus that opportunities should become objectives. The development of such a consensus will acquire a new appreciation of how policy agendas are established.

TOWARD POLICY

Policy follows, among other things, economic needs and social values. If one perceives the city as a set of socioeconomic systems, it is also necessary to lay out policy and research agendas that explicitly concern themselves with questions of social value, economic purpose, and political choice. These questions may not be perfectly resolved, but they do lead to answers associated with institutional and organizational development, which must be a primary component of urban development policy (Gappert, 1975).

In American industrial society the city has rarely been an acceptable symbol of a collective consumption ethic or of an integrative cultural style linking both pauper and prince. Instead, the industrial city has been more a symbol of the individual struggle for survival or private success, or of the ethnic competition for control of public resources. (For example, think of the novels by Theodore Dreiser or James Farrell.) The cities of an advanced industrial society—the future metropolis—will be primarily engaged in indirect, and partially abstract, transactional activities, and may be hungry for collective rites to offset social fluidity, economic transience, and electronic isolation. It may also be oriented to both material and nonmaterial standards and satisfactions integrated on a community and regional scale.

Although American society is not very accustomed to solving problems or resolving dilemmas by the use of what are called strategies, it is likely that the efforts to articulate a national urban policy will continue. The idea of strategy, partially borrowed from both the military and the foreign relations field, implies the development of a single, comprehensive, long-term plan to cope with some significant condition or opportunity. A strategy involves a set of basic assumptions about a style of action judged most appropriate to accomplish some major goal over a substantial period of time.

Given the assumption of an economic transformation, an overreaching urban strategy must in fact be an industrial strategy. Indeed, it should be noted that American rural strategies, developed after the United States was engaged in the Industrial Revolution, are in fact agricultural policies and have often been commodity-specific. Similarly, it is the acceleration of the communications revolution that is creating a need for an industry-specific industrial policy, which will provide for the more deliberate use of urban resources, especially its human capital in the form of both people and institutions (Drucker, 1982). (It is also possible to speak of an advanced industrial agriculture as in Bradshaw, 1979.)

To further suggest some possible dimensions of an emerging urban policy that is also an advanced industrial policy, it is necessary (1) to review current efforts to develop a national urban policy, and (2) to consider several alternative policy paradigms.

NATIONAL URBAN POLICY

In early 1981, shortly after the release of the reports of the President's Commission for a National Agenda for the Eighties (1980), the National Research Council (NRC) established the Committee on National Urban Policy to conduct a four-year study. According to their first report:

> Urban policy resists precise definition. Historically, it has been considered (among other things) a component of subnational regional policy, a euphemism for much of domestic social policy; an umbrella term describing policies to deal with places; and as programs specifically addressed to the physical, fiscal and social afflictions of central cities [National Research Council, 1982].

A decade earlier Moynihan (1970), just prior to joining the Nixon administration, had enunciated ten policy heuristics that he called a national urban policy. But the policy history of the intervening years indicates that the quest for a single strategic perspective on urban public policy may turn out to be a fool's errand.

The initial NRC study has, however, identified five issues that it believes will "shape the agenda for national urban policies over the next decade and perhaps beyond." These issues are:

—The mobility, or lack of mobility, of people and jobs, including the impact of legal and illegal immigrants on cities.

—The question of whether the nation is developing an urban underclass, especially along racial lines, that is being perpetually shut out of the mainstream economy.

—The federalism debate and issues surrounding whether states are sufficiently capable of handling the new authority and responsibility being given to them.

—Relations between the private and public sector, especially the question of capital formation for urban investment in older central cities.

—The decline of the urban infrastructure of roads, bridges, and water and sewer lines.

These issues represent significant problems faced by cities and society today. The new articulation of the urban underclass problem, with note being taken of the increase in the relative concentration of poverty among blacks and Hispanics in urban areas, is of special interest. The report deplores "the apparent persistence of this condition despite efforts to expand opportunities for the structurally unemployed."

But, given the post-affluent framework previously outlined, these efforts were undertaken when the mainstream economy was experiencing several dislocations, new constraints, and sectoral contractions in some of its most important industrial sectors (steel, auto, rubber, etc.). The unfortunate reality is that the national efforts to shrink the size of the underclass were undertaken at the same time that (1) the economy had to absorb a substantial number of young workers, and (2) economic constraints and dislocations were dramatically increasing the number of "frictionally unemployed " (see Blair, this volume).

At this point the attraction of some kind of supply-side economics is rooted in the reality that any focus of national policy must be on expanding the aggregate structure of opportunities across the entire mainstream economy.

As the post-affluent transition continues, national policy efforts in any sphere can either (1) contribute to further economic contraction and decline, (2) emphasize purely economic issues by focusing on reorientations of life styles and working arrangements, or (3) accelerate the structural shifts necessary to guide the economy and society to an advanced industrial status in the emerging global economic system.

The danger, however, is that efforts will be made to bring back, at great expense, the old urban industrial order—to engage in fiscal raindances in hopes that buffalo will return—and to rebuild an urban infrastructure that was created 100 years ago when the first Industrial Revolution was just under way.

A commitment to a genuine national urban policy should follow from an appreciation of the dynamics and opportunities of an advanced industrial society. But since the advanced industrial society is not yet—and may not ever be—an empirical reality, a neourban policy, constructed from the remnants of New Frontier-Great Society programs (such as HUD's highly regarded Urban Development Action Grant [UDAG] program), is more likely to be advocated and accepted.

Table 1.1 provides two tentative alternative policy models. The first suggests the elements for a neourban national policy and the second provides analoguous policy elements based upon an advanced industrial framework. At this point the elements are only illustrative and even simplisitc; the two models are not incompatible but complementary. The notion of a high-tech urban policy needs to be elaborated further during the next half decade. A significant issue related to its elaboration is the recognition that in both sunbelt and snowbelt cities, skilled-labor shortages are increasingly being filled with immigrants for whom English is a second or third language (according to the New York *Times,* May 16, 1982, American hospitals recruit 6000 Filipino nurses annually). The resocialization of members of the semiskilled blue-collar working class to knowledge-intensive employment may be the single most significant domestic issue of the 1980s. (A less significant issue perhaps is new patterns of American emmigration in selected occupations as the global economy evolves further.) At this point it can be useful to turn to several policy paradigms as a way of suggesting how new urban programs and issues may be developed.

Table 1.1

Alternative Policy Models

A Neo-Urban National Policy	An Advanced Industrial Policy
(1) Deplore and resist the consequences of unanticipated economic mobility.	(1) Anticipate changing economic demands for skilled labor and its location.
(2) Expand efforts to shrink the urban underclass.	(2) Facilitate working-class adjustment to new labor-market requirements.
(3) Retain direct federal assistance to cities.	(3) Focus federal attention on urban linkages to foreign economic policies and expansion of global manufacturing system.
(4) Provide substantial incentives and tax and regulation relief to targeted private sector urban investments.	(4) Provide substantial incentive for industrial investments in new, national technological prioritiets.
(5) Rebuild and repair urban infrastructure.	(5) Dramatic expansion of investment in human capital in urban area technical colleges, urban program universities, retraining vouchers, etc.)

POLICY PARADIGMS

It may be that "paradigm" is one of the most overused words of this decade. Ever since Kuhn wrote the *Structure of Scientific Revolutions* (1970), social theorists and especially futurists have been using the concept to juxtapose the different assumptions that guide popular thinking in alternative and sometimes opposite directions. For example,

> A paradigm consists of particular theories, techniques, beliefs, and values shared by the practitioners of a science. More specifically, it includes a particular set of symbolic generalizations, belief in the efficacy of certain well-defined models of reality, and shared values among practitioners as to the criteria for effective or adequate solutions.

> Paradigms are more than models. Different models may be contained within a paradigm. Paradigms reflect more of a spirit of inquiry, a kind of world view through which beliefs and reality experiences are generalized into theory. Lacking any acceptable urban theory which is more than an expedient assemblage of propositions borrowed from sociology, political science, and other disciplines, the urbanist, be he scholar or practitioner, has to be more self-conscious of the biases which govern the way in which he formulates the problem and generates and selects "solutions" or "explanations."

> There are perhaps four paradigms which can be used to characterize (or caricature) different approaches to understanding and interpreting urban systems for the purposes of strategic policy development [Gappert and Rose, 1976].

These paradigms are illustrated in Figure 1.3. The technocratic paradigm is organized around concerns of rationality and efficiency (Do the buses run on time?), and contributes to cities of intentional change and cities of efficiency. This perspective has a concern with planning models directed at the more efficient use of urban resources (polynucleated systems) as well as at building entire new systems (Harbor Place in Baltimore).

The pragmatic humanist paradigm is more concerned with the quality of human behavior (Was the bus driver polite?). This perspective contributes to the social efficiency of cities as well as to cities with good political machines. This perspective will be necessary to ensure that people will have the opportunity to both define and express their own needs and problems and to contribute to the solutions. Similar skills will be necessary to insure that new regional alliances are formed and that new immigrant communities become part of a stable

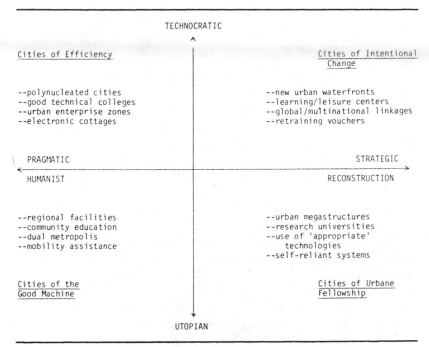

FIGURE 1.3   Paradigms for Urban Policy

but dual metropolis (see Bloomberg and Martínez-Sandoval, this volume).

The utopian paradigm is characterized by attempts to affirm a well-defined, perfect future city (We won't need buses when the beautiful subway is completed). In American society utopian outcomes are not just generated by poets and philosophers but by hard-headed people who can get things done. The ideal of managed progress—which asserts that with the proper combination of skill, technology, resources, and good intentions, things will come out better—has always been an element of urban industrial culture. The newest expressions of this perspective have a particular attraction to the so-called appropriate technologies associated with solar energy, urban gardening, and other, more self-reliant systems. The utopian perspective also has an underlying faith in the perfectability of urban man, and in the ultimate achievement of a new kind of neighborhood. (There is also a dystopian paradigm that cities are irredeemable sites of strange and undesirable behavior, see Merry, this volume).

The fourth paradigmatic category reflects a strategic reconstruction perspective concerned with guiding or directing the evolution of the urban community within a holistic framework. The emphasis is on the design of long-range, larger-scale programs, targeted to the exploitation of new opportunities (Will the subway link the central business district to the new jetport?). This strategic perspective is concerned with the qualities of institutional vitality necessary to plan cities of intentional change, as well as with the more precise identification of the nature of new opportunities generated by new technologies and global developments.

American society is not undersupplied with policy proposals. It is perhaps less well supplied with research in support of policy development, and it is certainly short of the strategic formulations out of which an urban consensus can arise. The four paradigm catagories can be used to suggest ways in which policies can be added up to support different strategies related to a new and sharper sense of national purpose. Do we want to choose between drifting into System X, evolving into System Y, or designing a new System Z?

If we are beginning to develop into some kind of knowledge-driven economy, what is the role of the great urban and metropolitan centers? There is not a proven recipe for instant cities. As our society continues through a post-affluent transition, it lacks a reliable, predictive evolutionary model of change. Economic goals and opportunities are changing, and the economic means to achieve the social goals associated with the dreams of affluence in the 1950s and 1960s are no longer easily available.

Each of the paradigms can be used to support the design of new and enduring urban goals and values. Each perspective needs to be represented as plans for urban development appropriate to the possibilities of the twenty-first century are generated. In a time of scarce resources the efficiency imperatives of the technocratic paradigm cannot be ignored. The utopian paradigm is needed to generate some practical but bold new fantasies that can inspire communities to new efforts in the midst of the transitional stress. The pragmatic humanist perspective is clearly needed to develop proposals for more humane relationships in both our workplaces and neighborhoods. Together, the four perspectives can provide a commitment to an urbane civilization which can support networks of social relationships in an atmosphere of pluralistic tolerance, humane subsistence, and, dare we say it? technological progress.

REFERENCES

BELL, D. (1973) The Coming of Post-Industrial Society. New York: Basic Books.

BLAIR, J. P. and G. GAPPERT (1976) The Problems and Consequences of a Slow/No-Growth Economy. Washington, DC: Joint Economic Committee.

BRADSHAW, T. K. (1980) California as a Post-Industrial Society. Institute of Governmental Studies, University of California, Berkeley.

——— (1979) Rural Communities in Advanced Industrial Societies.

——— and E. BLAKELY (1979) Rural Communities in Advanced Industrial Society: Development and Developers. New York: Praeger.

DRUCKER, P. (1982) "The shape of industry to come." Industry Week January 11.

——— (1980) Managing in Turbulent Times. New York: Harper & Row.

FERGUSON, M. (1980) The Aquarian Conspiracy, Personal and Social Transformation in the 1980s. Los Angeles: Tarcher.

GAPPERT, G. (1979) Post Affluent America. New York: Franklin Watts.

——— (1973) "The future of economic inequality and the planning of urban services." American Institute of Planners' Journal (May).

——— and H. M. ROSE (1976) The Social Economy of Cities. Beverly Hills, CA: Sage.

GRAY, D. D. (1982) "The technological barnyard." Bulletin of the World Future Society (April).

HAMRIN, R. (1980) Managing Growth in the 1980s. New York: Praeger.

HIRSCH, F. (1978) Social Limits to Growth. Cambridge: Harvard University Press.

HIRSCHHORN, L. (1982) Beyond Mechanization: Flexibility and the Theory of Post-Industrial Technology. Cambridge: MIT Press.

JONES, L. Y. (1980) Great Expectations, America and the Baby Boom Generation. New York: Ballantine.

JOHNSON, W. G. (1979) Muddling Toward Frugality. Boulder, CO: Shambhala.

KNIGHT, R. V. (1980) The Region's Economy: Transition to What? Cleveland State University.

KUHN, T. S. (1970) The Structure of Scientific Revolutions. Chicago: University of Chicago Press.

LINDEN, S. B. (1971) The Harried Leisure Class. New York: Columbia University Press.

LIFTON, R. J. (1976) The Life of the Self. New York: Simon & Schuster.

MARIEN, M. (1973) "Daniel Bell and the end of normal science." The Futurist (December).

MASUDA, Y. (1981) The Information Society as Post Industrial Society. Washington, DC: World Future Society.

MENSCH, G. (1982) Stalemate in Technology. Cambridge: Ballinger.

MOYNIHAN, D. P. (1970) Maximum Feasible Misunderstanding. New York: Free Press.

National Research Council (1982) Critical Issues for National Urban Policy. Washington, DC: National Academy Press.

NIEBUHR, H. A Renewal Strategy for Higher Education. Temple University, Philadelphia.

NEVIN, J. (1982) Slide presentation. Akron Roundtable.

OGILVY, J. (1979) Many Dimensional Man, Decentralizing Self, Society and the Sacred. New York: Harper & Row.

OUCHI, W. G. (1981) Theory Z. New York: Avon.

PHILLIPS, K. (n.d.) Post Conservative America. (unpublished)

President's Commission for a National Agenda for the Eighties (1980) Urban America in the Eighties, Perspectives and Prospects. Washington, DC: Government Printing Office.

SCHWARTZ, P., J. OGILVY, and P. HAWKEN (1982) Seven Tomorrows, Toward a Voluntary History. New York: Bantam.

TOFFLER, A. (1981) The Third Wave. New York: Bantam.

YANKELOVICH, D. (1981) "A world turned upside down." Psychology Today (April).

YOUNG, M. and P. WILLMOTT (1973) The Symmetrical Family. New York: Pantheon.

Part I

Urban Changes:
What Pattern Is This Place?

☐ ANY PERSPECTIVE ON THE FUTURE must begin with a historical orientation. Good futurists, like most historians, have to have a substantial appreciation of the significance of time in the affairs of our societies, cultures, and cities.

The patterns of urban change and development must be described and understood, both for the national system of cities and for individual urban places. Cities grow as centers of employment opportunities but develop as sources of amenities, cultural resources, and cosmopolitan opportunities that transcend the purely material.

Richardson reviews the emergence of the American urban system of cities that began around 1840 and continues to the present. Knight focuses on the future development of the industrial city, attempting to outline the logic of its evolution into a more complex and cultured type of urban place. Shostak, using materials that were developed for the tricentennial celebrations in Philadelphia (1682-1982), presents seven scenarios of urban change that have relevance for most cities today. He reminds us of the importance of choice—collective and individual—in the achievement of urban progress, a desirable element in any pattern of change and development.

2

The Evolving Dynamics of American Urban Development

JAMES F. RICHARDSON

☐ HISTORIAN RICHARD HOFSTADTER once wrote that the United States was born on a farm and moved to the city. As a high-level generalization this statement holds for much but not all of the national history. Until recent decades each census would show both an increase in the number of urban places and a higher proportion of the population living within them. In the early nineteenth century, American cities served primarily as commercial centers funneling primary products to external markets and receiving finished goods, capital, and immigrant labor in return. Until at least the 1830s the American economy remained a colonial one without a developed national urban system. Its principal cities looked to Europe for ideas, institutional models, and financial leadership, a relationship that remained throughout the nineteenth century, although the balance shifted somewhat as American cities acquired greater economic and institutional maturity. Before 1840 most manufacturing in the United States was carried on outside of cities; industrial activity within the cities either serviced the local market (baking of bread and brewing of beer) or was ancillary to commercial functions (boat and ship building). Cities grew before 1840, both in number and size, but at a slower pace than would be the case in the next two decades (Pred, 1966, 1973).

Between 1840 and 1860 the United States moved from a set of only loosely linked regional economies into a national urban system dominated by New York City. A case in point is the banking industry. Banks all over the country maintained correspondent relationships with New York banks because of that city's financial and economic supremacy (Conzen, 1975). Internally, New York, Philadelphia, and other cities were transformed by industrial growth and heavy immigration. Foreign and domestic investment in railroads and the other transport devices provided necessary links in the national urban system. Industrial growth provided job opportunities for immigrants whose places of origin had grown rapidly in population without a corresponding increase in economic activity. Southern Ireland is the classic example (Thomas, 1954).

The volume of immigration illustrates the magnitude of growth and the extent of change. Between 1776 and 1820 the best estimate of the total number of European immigrants who settled in the United States is about 250,000 people. For the single year 1854, the peak of antebellum immigration, the total exceeded 450,000. During the 1850s nearly three million immigrants arrived. More than one out of seven persons living in the United States in 1860 had been born outside of its borders, a proportion about the same as the 1910 peak of the preworld War I influx of labor. In 1860 the foreign-born populations of northern cities ranged anywhere from one-fourth to one-half of total population, giving them a distinct Hibernian and Germanic flavor (Jones, 1960: 117-118).

In Philadelphia, which due to the Philadelphia Social History Project has received the most detailed scrutiny, most male immigrants worked as laborers or at industrial tasks ranging from traditional crafts to operatives in increasingly mechanized factories. From the 1840s on, industrialists chose urban locations because they provided the most ready access to information, transportation, labor, and ancillary enterprises. The use of steam power in manufacturing meant that factories no longer had to be located at water power sites. The cost of transporting coal was such that cities with river or canal connections had an important locational advantage. Within cities people chose their residential locations on the basis of their employment. All but the most affluent walked to work and only a few lived more than a mile from their jobs. In contrast to the modern metropolis, which can be conceived of as a single labor market, nineteenth century Philadelphia consisted of a number of labor markets. The desire to live close to work was so strong that a change of place of work could easily mean a change of residence. Correspondingly, most employers probably were limited in their recruiting to perhaps a one mile radius from their businesses. The location of industrial activity was thus the biggest single influence upon the city's land use patterns (see Greenberg; Hershberg et al.; and Burstein in Hershberg, 1981).

The location of a particular industry within a city in turn depended upon the nature of its market and the amount of land it needed. As geographer Allen Pred has indicated for New York, printing and publishing required close access to political and business leaders but little space; therefore such firms located in the core. Space-extensive activities such as shipbuilding were forced away from prime areas, although yards still had to be along the waterfront (Pred, 1966). The famous Baldwin locomotive works moved three times during the nineteenth century, each time further away from the center of Philadelphia (Jackson, 1973). When an industry chose a peripheral rather than a central location, so too did most of its employees, although some workers reverse-commuted, central locations meant more choices in case they lost a job and had to find another, and were more convenient for other working members of the family. In the last century these additional workers were more likely to be children than wives. The Philadelphia data show only a limited number of married women working for pay outside the home. In contrast, working-class children began contributing to family income at an early age—before age fourteen—and usually continued to do so into their twenties. One sample of males showed an average of seven years between entrance into the full-time labor force and leaving home. Since working class incomes peaked in the 30s and many men died at early ages, children's earnings were important components of total family income (see chapters by Haines, Goldin, and Modell et al. in Hershberg, 1981).

Up to about 1840 urbanization and industrialization were not necessarily related; it was possible to have one without the other. Even into the early twentieth century some cities developed as regional centers with minimal industrial components. In 1910 Los Angeles had a lower figure for value added by manufacture than did Akron, a city with less than one-fourth the population of Los Angeles. However, for most large American cities between 1850 and 1920 the phrase "urban industrial society" came ever closer to approximating reality. It was industrial growth that fueled urban growth and industrial location that determined patterns of urbanization both on the macro level of the position of an individual city within the national urban system and the micro level of land use within a given city (Pred, 1966; Ward, 1971; Fogelson, 1967).

During the late nineteenth and early twentieth century, urban and industrial growth was concentrated in the northeastern quadrant of the nation. The urban-industrial core, bound on an east-west axis by New York and St. Louis, was served by a thick belt of railroads, making it feasible for final product manufacturers in Pittsburgh to import intermediate goods from Cleveland, and for final product producers in Cleveland to buy semifinished goods from Pittsburgh. Railroad rate policies sanctioned by the Interstate Commerce Commission (ICC) gave preferential treatment to enterprises located within this core (known in railroad parlance as "Official" territory). By 1900 New York and Chicago were linked by telephone as well as telegraph, and cities such as Cleveland pushed for the westward extension of the Eastern Time Zone so that they would be on the same time as New York. An efficient mail service often managed next-day delivery between major cities located within the core (Potter, 1965; Ward, 1971).

During these decades cities became bigger both in population and land area, more diversified in their economic and occupational structures, and more specialized in land use. By 1900 virtually every important American city had been founded; almost all new urban creations in the twentieth century were satellites of existing cities. The modern central business district (CBD) with its office buildings, department stores, and theaters reached maturity by the 1880s, although subsequent decades experienced much tearing down and rebuilding on a larger scale to accommodate a higher level of economic activity. CBD expansion reduced the housing available in the urban core, including some of the mansions of the wealthy, which went down to make room for profit-making uses. Some of these mansions had such a short life that they could be considered as mobile homes for the rich. Red light districts, larger in area than most Americans would now believe possible, shared some of the downtown space. Public transit routes converged on the CBD and brought in office workers, shoppers, and entertainment seekers from peripheral areas (Goldfield and Brownell, 1979: 222-240; Chudacoff, 1981: 89-91).

These new residential areas growing on the periphery of the city or beyond its borders were built to high standards and were therefore accessible only to those of above-average income. Unlike preindustrial cities, whether in medieval Europe or the contemporary third world where the elite preempt the center and the poor throw up shacks on the periphery, American cities from the late nineteenth century for the most part have had a poor core and an affluent ring. This is not to say that some central residential areas were not upper income—a case in point is Boston's Beacon Hill—or that there were not peripheral working-class areas clustered around factories and railroad yards. Nor were income and job location the only factors influencing residential location; people with children were often ready to trade off a longer journey to work for more space. (This brings to mind Robert Benchley's comment that in America there are two kinds of travel, first class, and with children.) But the basic equation of inner city with lower income was in evidence by the 1880s (Downs, 1973: 14).

Many suburban areas throughout the nineteenth century became part of their central cities, either voluntarily to take advantage of city services such as water and police protection, or involuntarily because of forcible annexation sanctioned by state legislatures. Some of the more famous annexations in the nineteenth century include the expansion of Philadelphia in 1854 from two square miles to all of Philadelphia County—almost 130 square miles, Chicago's absorption of much of the present-day South Side in 1889, and the creation of Greater New York, which took effect in 1898. During the twentieth century these and other northern cities have found it increasingly difficult to annex territory. Suburbs could finance their own services, sometimes with the city's unwitting help, and resist annexation movements in state legislatures. The spread of state constitutional provisions guaranteeing municipalities "Home Rule" made it even easier for larger suburbs to maintain their autonomy (Jackson, 1973).

By the 1920s the movement of the affluent beyond the city's borders was beginning to have serious effects upon cities such as Cleveland. The suburban migration removed money and talent, and the movers and shakers of the community had a more narrowly focused view of their obligations toward the community than had their predecessors a generation earlier. Warner (1968) has argued that a crucial turning point for American cities came in the mid-nineteenth century when the economic activities of their elites transcended municipal boundaries. In the 1820s American cities had a unified elite. The same group dominated economic, political, and nongovernmental institutional life. By the middle decades of the century a specialization and diffusion of leadership had begun. The same names were no longer found both on a city's business firms' letterheads and the city council. The combination of a democratized polity, with its greater number of voters and appeals to the myth of the self-made man, and a shift of attention on the part of the elite to wider horizons, left municipal government to second and third echelons, including the professional politicians, a species suspect almost by definition. Men with national economic interests and who, by the late nineteenth century, considered themselves part of the national upper class, had less time, attention, and concern for the particular city they lived in (Warner, 1968).

It seems to me that another such turning point came when the elite moved their residences beyond municipal borders and became less concerned about the quality of the city's school system or the physical environment of its residential neighborhoods. Cities in the 1920s spent more money on these matters than they had before World War I; still, there was a qualitative difference in the economic leaders' views that would have long-term quantitative consequences. Before 1910 most of Cleveland's business and professional leadership lived within the city's corporate boundaries. By 1925 the situation was reversed and most of the affluent lived in the suburbs. The census of 1930 showed that the suburbs collectively were of higher socioeconomic status than the city. A close reading of the material from Chambers of Commerce and other such groups indicates a greater distancing of Cleveland's economic and professional leaders from the "cosmopolitan" population of the city then had been the case before 1915. Old-stock members of the elite had disdain for the representatives of new immigrant groups and blacks coming to the fore politically in the 1920s. When organizations like the Citizens League and others spoke on Cleveland government and politics, members of council responded that residents of Shaker Heights had no right to tell residents of Cleveland how to conduct their affairs. (Cleveland Chamber of Commerce Minutes, 1900-1930).

The traditional view of local government assumed that all of people's activities would be carried on within a single jurisdiction. They would work, live, play, and educate their children within the same taxing and service area. Decisions made in their business lives would have a direct impact upon their living environment; if there were negative consequences from pollution they would suffer them along with everyone else. The physical separation of work and residence severed these connections. If city school systems deteriorated, suburbanites would not suffer so long as they could maintain their good schools.

On the broader measure, the constantly increasing scale and centralization of decision making over the last century means that fewer communities than ever are masters of their own economic destiny. To be sure, in the nineteenth century the fate of New York commercial houses could depend upon decisions made in London, just as Kansas City's future was contingent upon actions of men in Chicago, New York, and Boston. The rise of national corporations from the 1880s on took more decisions out of the market and put them under management. What Chandler (1977) has called the "visible hand" of management "replace[d] the invisible hand of Adam Smith's market forces.'' Corporate managers in New York or Chicago decided where their production facilities would be located, where to expand, and where to contract. The results could be boom or bust for communities hundreds, or even thousands of miles away from the decision makers (Hays, 1974).

The same centralization characteristic of economic activity also transformed political and social life. Ward schools, heavily influenced by local groups, gave way to city-wide school systems dominated by professionals. Police precinct commanders lost some of their autonomy to headquarters. Increasingly, Americans identified themselves more by function than locality. For middle-class Americans at least, the corporations they were with, and the organizations to which they belonged were the most important points of belonging and contact, not their neighbors. Increasingly, success came to be conceived of as climbing the organizational ladder with less and less attention paid to place.

In recent decades the scale has expanded even further as corporations have grown into multinationals with plants and distribution centers around the world. One can envision a scenario of American-dominated multinationals with healthy balance sheets and earnings records and a depressed American economy, especially in the central cities and older metropolitan areas. The picture is not all bleak. Portions at least of those older cities that retain a strong base in advanced services can survive and perhaps even flourish. Cities have traditionally been the physical setting for face-to-face communication among people who fill different but complementary roles. Electronic communication may obviate the need for some direct meetings, but there are still many situations that require eyeball-to-eyeball interaction.

The result is a paradoxical one. The core of a city like Philadelphia gives off a sense of vitality and renewal while the life blood drains out of the city's industrial areas. the visitor who spends all of his or her time downtown thinks the city is healthy; the blue-collar worker whose job has gone with a plant closing has a different opinion.

Lane (1979) describes the situation of blacks who have migrated to Philadelphia in recent decades as having been "piped aboard a sinking ship." Blacks and Hispanics, whom economics and hostility confine to central cities and some older suburbs, find that these areas no longer provide a large number of entry-level blue collar jobs. Blacks' current situation resembles more closely their own previous history in the city than it does that of European immigrants of earlier periods. Philadelphia in 1860 had 22,000 blacks, about 4% of its total population, compared to the 640,000 persons and 38% of 1980 (Bureau of the Census, 1981). The earlier group was also residentially segregated and economically deprived, just as the majority of the city's black population is in the 1980s (Hershberg, 1981).

In the course of the twentieth century blacks have moved from the least- to the most-urbanized group in the nation's population. For the first half of the century whites were also urbanizing, although their settlement patterns were more outer city and suburban, in contrast to the black concentration in the inner city. In the last three decades this trend toward concentration has been reversed, to the point where Berry (1973: 27-73) has spoken of counter-urbanization. By 1960 suburbanites outnumbered central city dwellers. By the 1970s settlement patterns changed to the point where nonmetropolitan areas grew more rapidly than Standard Metropolitan Statistical Areas (SMSAs). Cleveland has lost one-third of its 1950 population, and portions of the city have lower densities than the upper-income suburb of Pepper Pike. Older housing has been abandoned and torn down, and nothing has been built in its place. Declining densities do not necessarily reduce social problems, for many of the individual dwelling units that remain are just as crowded as they ever were.

The key variable in whether an area is flourishing or declining is not age, but the level of capital investment. In most cases it has been more profitable to invest in newer areas than in older, and, as previously noted, many of the newer industrial areas are located outside the United States (Cohen, 1977).

What, if anything, can be concluded from this survey of American urban history? Perhaps the most banal but still important conclusion is that the city of the future will not be like the industrial city of 1850 to 1950, and there is not much point in planners and policy makers trying to bring it back. Since less and less of the labor force is resource-bound, and entrepreneurs are freer than ever before in choosing where they will conduct their operations, it seems as if the cities that will survive and flourish are those that maintain and develop the highest level of amenities. In the past people came to cities like Akron and Detroit because they knew they could find work there; that consideration outweighed any consideration of the kinds of places they wanted to live in. Now, and for the foreseeable future, it looks as if those cities that people want to live in will be those that generate the greatest employment opportunities. There will always be special cases, Houston's energy-related growth of the 1970s for example, but if the past is any guide, the future may belong to the urbane.

REFERENCES

BERRY, B.J.L. (1973) The Human Consequences of Urbanization. New York: St. Martin's Press.

BURSTEIN, A. (1981) "Immigrants and residential mobility: the Irish and Germans in Philadelphia, 1850-1880," in T. Hershberg (ed.) Philadelphia: Work, Space, Family and Group Experiences in the 19th Century. New York: Oxford University Press.

CHANDLER, A. D., Jr. (1977) The Visible Hand. Cambridge: Harvard University Press.

CHUDACOFF, H. (1981) The Evolution of American Urban Society. Fnglewood Cliffs, NJ: Prentice-Hall.

Cleveland Chamber of Commerce (1900-1930) Minutes. Greater Cleveland Growth Association Papers, Western Reserve Historical Society, Cleveland, OH.

COHEN, R. B. (1977) "Multinational corporations, international finance, and the Sunbelt," pp. 211-226 in D. C. Perry and A. J. Watkins (eds.) The Rise of the Sunbelt Cities. Beverly Hills, CA: Sage.

CONZEN, M. P. (1975) "Capital flows and developing urban hierarchy: state bank capital in Wisconsin, 1895-1854." Economic Geography 51 (October): 321-338.

DOWNS, A. (1973) Opening up the Suburbs: An Urban Strategy for America. New Haven, CT: Yale University Press.

FOGELSON, R. (1967) The Fragmented Metropolis: Los Angeles 1850-1930. Cambridge, MA: Harvard University Press.

GLAAB, C. N. (1962) Kansas City and the Railroads. Madison: State Historical Society of Wisconsin.

GOLDFIELD, D. R. and B. A. BROWNELL (1979) Urban America: From Downtown to No Town. Boston: Houghton Mifflin.

GOLDIN, C. (1981) "Family strategies and the family economy in the late 19th century: the role of secondary workers," in T. Hershberg (ed.) Philadelphia: Work, Space, Family and Group Experiences in the 19th Century. New York: Oxford University Press.

GREENBERG, S. W. (1981) "Industrial locations and ethnic residential patterns in an industrializing city: Philadelphia 1880," in T. Hershberg (ed.) Philadelphia: Work, Space, Family and Group Experiences in the 19th Century. New York: Oxford University Press.

HAINES, M. R. (1981) "Poverty, economic stress, and the family in a late 19th-century American city: whites in Philadelphia, 1880," in T. Hershberg (ed.) Philadelphia: Work, Space, Family and Group Experiences in the 19th Century. New York: Oxford University Press.

HERSHBERG, T. [ed.] (1981) Philadelphia: Work, Space, Family and Group Experiences in the 19th Century. New York: Oxford University Press.

HERSHBERG, T. et al. (1981) "A tale of three cities: blacks, immigrants and opportunity in Philadelphia 1850-1880, 1930, 1970," in T. Hershberg (ed.) Philadelphia: Work, Space, Family and Group Experiences in the 19th Century. New York: Oxford University Press.

——— (1981) "The 'journey-to-work': an empirical investigation of work, residence and transportation, Philadelphia 1850-1880," in T. Hershberg (ed.) Philadelphia: Work, Space, Family and Group Experiences in the 19th Century. New York: Oxford University Press.

JACKSON, K. T. (1973) "The crabgrass frontier: 150 years of suburban growth in America," in R. A. Mohl and J. F. Richardson (eds.) The Urban Experience: Themes in American History. Belmont, CA: Wadsworth.

JONES, M. A. (1960) American Immigration. Chicago: University of Chicago Press.

MODELL, J., F. J. FURSTENBERG, Jr., and T. HERSHBERG (1981) "Social change and transitions to adulthood in historical perspective," in T. Hershberg (ed.) Philadelphia: Work, Space, Family and Group Experiences in the 19th Century. New York: Oxford University Press.

POTTER, D. M. (1965) "The historical development of eastern-southern freight rate relationships," in R. Abrams and L. Levine (eds.)The Shaping of 20th-Century America. Boston: Little, Brown.

PRED, A. R. (1973) Urban Growth and the Circulation of Information: The U.S. System of Cities 1840-1970. Cambridge, MA: Harvard University Press.

———(1966) The Spatial Dynamics of U.S. Urban Industrial Growth, 1800-1914. Cambridge: MIT Press.

THOMAS, B. (1954) Migration and Economic Growth. Cambridge, England: Cambridge University Press.

WARD, D. (1971) Cities and Immigrants. New York: Oxford University Press.

WARNER, S. B., Jr. (1968) The Private City. Philadelphia: University of Pennsylvania Press.

3

City Development in Advanced Industrial Societies

RICHARD V. KNIGHT

☐ AS THE WEALTH CREATED in the twentieth century is reinvested in the twenty-first, the economy of cities will be transformed, new values will be articulated, new industries will emerge and older ones will adjust to the realities of global markets, the nature of work will change, and city life will reap the benefits from lessons learned over a century of urban living. In short, the functions, fortunes, and form of cities will change in a very significant manner. And new types of cities will emerge as a result of the second Industrial Revolution.

The forces that underlie the development of cities will be very different in the twenty-first century. The role played by natural resources and locational advantages will be eclipsed by human resources and cultural advantages. As these man-made resources, which are created by communities themselves, become more important, city development will become more intentional. Moreover, as design standards are upgraded to meet the requirements of the human and cultural resources, planning will become more forward-looking, cities will become more cosmopolitan and increasingly desirable as places to live.

DIFFERENT FUTURES, DIFFERENT PATHS, DIFFERENT CITIES

Cities will, of course, follow different paths. Some may continue to take their future for granted and simply react to ever-changing external forces. They will be buffeted by all kinds of foreseeable and some unforeseeable storms. Others will become more autonomous and, with an increased understanding of the forces that underlie their development, will design new futures. Those that design futures that are both feasible and desirable will lead the way. Considerable foresight and leadership will be required to create the consensus and mobilize the resources required to position the cities in the emerging international order. But cities that gain control over their destinies will prosper. Accidental cities that are not able to nurture a sense of community or a sufficient collective commitment to a concerted course of action or development strategy are unlikely to have the resiliency required during the turbulent times that lie ahead. The outlook for cities that are drifting is not promising. As historians have noted, history is a graveyard of cities. Do-nothing cities, those lacking foresight and initiative, will become resigned to an erosion of position and power, at least in relative terms.

Cities, their institutions, their economies, their neighborhoods, and their citizens are, by nature, continuously in flux. In the present century growth has primarily been a process of accommodating change brought about by the Industrial Revolution and urbanization. But as the new international order of cities is established, and as the movement of people, capital, resources, ideas, and information becomes freer, the actual migration of people is likely to become more selective and will diminish. The growth of cities will stabilize as the roles they play in their respective regions and in the global economy become more clearly differentiated. City development in the industrialized nations in the next century will take the form of qualitative changes, rather than that of expansion of people and territory as occurred in the first two centuries of the industrial and urban revolution. City development in countries that are becoming urbanized and industrialized will continue to occur very radically.

The form that cities take both in terms of opportunity and urban design will become more distinct, and citizens will become more sophisticated contributors to, and consumers of, cities. This is not to suggest that people will not have mobility (travel and tourism will undoubtedly be the largest industry in the next century) but rather that people will become more grounded. Citizenship may once again connote allegiance to and a sense of identity tied to a place as much as to a nation. This does not preclude the concept of a global village, such an option will certainly exist, but there will be a choice. One can live in the village and be wired to the world or live in a city in direct contact with civilization. As technology conquers distance, the city as a value will become stronger.

Why, in the face of all the problems that cities now confront, would one make a case for the resurgence of cities in the emerging international order? The reasoning is explained below; it hinges on the direction and form that advanced industrial society has been taking over the last several decades. Cities appear to be returning to their more traditional roles, roles they played before they were overrun and overwhelmed by the forces that fueled the Industrial Revolution and the concommitant rural-to-urban migration.

The industrial city is still in its formative stage of development. It is a very young city in the history of cities, for it is a child of the Industrial Revolution. But it is beginning to mature and thus to take its eventual form. The form that it takes will be difficult to anticipate because the role of the industrial city in the global economy is still not clear. It is clear that the eventual form of the industrial city will not only be different from that of early industrial centers but its function will also be different from that taken by traditional cities. This distinction between an advanced industrial city and more traditional cities is very important. Unless such a distinction is made, it will be very difficult to advance our thinking and policies concerning city development and design.

THE CITY AND INDUSTRIAL DEVELOPMENT

New forces underlie industrial development and a new conceptual framework is needed that will account for this type of growth and the way it affects city development generally. There is, at present, widespread concern about the waning of traditional manufacturing activities, plant closings, and the loss of blue-collar jobs, but there is very little awareness or appreciation of the new forms that industrial activity is taking and how the transition is occuring. Opportunities in advanced industrial activities are expanding, but they do not fit into the conventional modes of thought and are thus rarely considered. Once these new opportunities have been conceptualized and once an appropriate conceptual framework is elaborated, these activities will be better understood, policies will be formulated that build on these new opportunities, and confidence in the future of industrial cities will gradually be restored.

First, it should be noted that the concept of a city is not usually very well defined. Certainly, designation of city status to a local government by a state government does not mean that such a place is a city in the broader historical sense of the term. Similarly, just because a city is unable to expand its territory by annexing the surrounding areas that its residents are moving to, does not mean that a city is in decline. In fact, as a city develops, the city's area of influence usually expands as institutions based in the city extend the region they serve by upgrading the quality of services provided. As these regional organizations grow, the structures that house them, such as government offices, hospitals, terminals, universities, sports arenas and cultural facilities, access streets, and parking, have to be enlarged, a process that usually involves a change in land use. Naturally, population declines as residential areas are converted to nonresidential uses. Thus, a decline of population may be anticipated in a developing city and should be viewed as part of the present growth process. Similarly, the fact that citizens have increased their skills and incomes and are taking up residence in newer and better planned suburbs outside of the municipal limits of the city should not be regarded as a decline of the city; to the contrary, it should be regarded as a sign of success. (We will return to the important issue of how to change the city growth dynamic so that the city may become a more cosmopolitan community by growing inward instead of outward.) Clearly, municipal boundaries lose their significance when development spills over into the suburbs and exurbs, or when development of nearby cities converges or overlaps. The concept of a city needs to be redefined.

DIFFERENT TYPES OF CITIES

Today, most cities need to be considered in the context of development occurring in their respective regions regardless of how municipal or even state boundaries are drawn. Many of the most politically balkanized cities are already bound together by a variety of metropolitan or regional (and sometimes multistate) authorities. A modern regionalized city is not structured as much by municipal services as by functional relationships such as commuting, broadcasting, and networks of private and non-profit organizations that have evolved and serve the region. Industrial cities tend to be more decentralized than traditional cities for a variety of reasons. To wit, great caution must be taken when comparing cities because many of the measures traditionally used to study the growth of cities, namely the expansion of population and territory, are no longer appropriate indicators of development.

The problem of how to define cities is not only a consequence of recent developments, it was also a problem in the past. There is no clear consensus on how to define a city but, if we are to clarify our thinking, we must at least attempt a definition. Perhaps the most helpful definition for our purposes is the traditional one, that cities are centers of civilizations; the centers of power that provide order to the civilizations over which they dominate. This is helpful because it acknowledges that the city is and always has been an elitist institution. It also suggests that the key to understanding a city lies in the nature of the power on which organizations that form the city's institutional base are founded. In a similar vein, the city has also aptly been described as the anvil of civilization; the place where new values are forged and administered for the civilization that uses them. This approach to defining cities will be used below because it yields important insights into the future of cities in an advanced industrial society.

Traditionally, cities were founded on what has generally been referred to as the moral order but during the last century or so a new type of city, the industrial city, has emerged on what may best be termed the technological order. Founded on different types of power, these two types of cities differ in important ways which, once recognized, help us understand the nature of development processes that will shape cities in the twenty-first century. Up until the Industrial Revolution there had been one predominant type of city, notably the imperial cities which presided over empires through a network of provincial or colonial capitals. This type of city can be traced back five or six millennia. After the fall of the Holy Roman Empire, however, commercial or mercantile cities were established; they evolved very slowly during the Middle Ages, and then began thriving, particularly during the Renaissance and in the mercantile and colonial eras.

Traditional cities were all essentially alike in that they were founded on the traditional or moral order. In the imperial city power was derived from control of land. Power was centrally vested in the emperor, the king, the pope, or the prince and was passed down to barons, bishops, and the governors who administered the territories in a hierarchical manner. Each city acted as the seat for the authorities responsible for governing a particular territory. Their essential function was to preserve the status quo.

The commercial city was founded on a new form of power that operated within the traditional moral order. Their power was based on liquid capital or exchange values. As trade grew, the merchant and middle classes grew and the cities gained in stature. Pirenne (1970) writes about how these cities gradually gained power and autonomy in the Middle Ages. It was not, however, until they had grown sufficiently strong to wrest control from the bishops and barons that these commercial towns became autonomous and gained the status of cities. Only then were they able to loosen the restrictions on trade, expand their markets, and gain control over their destiny. The existence of a large, dominant middle class and local autonomy are considered the critical attributes of cities even today.

Up until the advent of gunpowder, double-entry bookkeeping, commercial drafts, the joint stock company, and limited-liability investments, commercial or mercantile cities were truly a community effort and took the form of a work of art. These cities were clearly defined; they were compact settlements surrounded by walls and entrance was controlled through a limited number of gates. Citizens took great pride in their cities and defended them with their lives. The capital that they accumulated was invested locally, not in the short-term investments that are so prevalent today, but in structures that were built to last forever. There was, at that time in history, a real sense of place. Trade increased and commerce quickened. Guilds were organized, new forms of power were structured, an educated middle class emerged, and the pool of liquid capital or credit available for new ventures grew. The stage was set for the Industrial Revolution.

Then, with the advent of the scientific and industrial revolutions, new types of opportunities were created and still another kind of power began to take form. Eventually, this new type of power would provide the basis for building a new type of city—industrial cities, founded on the control of technology. During their early stages of formation, industrial cities evolved around the production of manufactured goods for export. Growth took the form of rapid population increases because technology was then in a primitive state of development and manufacturing was labor-intensive. Not much capital was required but as capital was accumulated and reinvested in plant and equipment more workers were needed. Labor was drawn to the mines, to mills, and to factories in places that provided locational advantages, and communities grew up rapidly around these new opportunities. As new products and processes were invented, or as new resources were discovered, new industries emerged, new profit-making enterprises were established, and new communities were formed. Some were company towns, others hosted many companies and a variety of industries; they took all sorts of shapes and forms. Most grew organically, a few were designed as planned communities; codes and restrictions were viewed as hampering growth.

Early industrial cities can best be described as historical accidents. Capital and labor were attracted to them and fortunes were made primarily in places that had natural advantages, that is, advantages based solely on their location. As competition for markets increased, access to raw materials, to energy (coal or water power), and markets became increasingly important. Entrepreneurs, capital, and labor gravitated to places with locational advantages. Manufacturing centers boomed and growth occurred in an untrammeled and a chaotic manner. As one architect observed, the industrial city was a shapeless mass formed by a thousand hands.

Early industrial centers that offered access to resources, energy, and markets did not have to do anything to encourage growth, it happened spontaneously. Manufacturing centers boomed; population grew and land values increased rapidly. The major challenge of the day was to build fast enough to absorb the growth. English rural workers, pushed off the land by the Enclosure Acts, by famine, or by the mechanization of agriculture, were pulled into the growing mill towns and manufacturing centers. As economic opportunity expanded in these newly formed urban settings, the great urbanization movement gained momentum. Work was shifted from the fields to the factories where tractors, farm implements, cars, trucks, and the like were manufactured to be used on the farms, thus displacing more workers from the fields. Instead of hoeing crop rows, workers now had to stand on the production line producing or assembling parts.

Capital formation was accelerated. Great fortunes were made as new resources and energy sources were discovered and exploited, and as the technology used to extract, transport, and process materials for manufacturing was advanced. Major inventions and innovations in products and production processes, such as the automobile and Henry Ford's assembly line, helped to establish new industries, new markets, and new manufacturing centers. Markets continued to grow as production costs were reduced through the economies of mass production and vertically integrated industrial complexes. The steel, electrical, petroleum, and chemical industries were formed. New technologies led to the combination of the factors of production in increasingly efficient manner. Profits were high but much of the capital created was used for plant expansion; production jobs increased commensurately, population increased, growth fed upon itself, and manufacturing centers boomed.

Wealth was created in several ways in the booming manufacturing centers, and most of it was reinvested locally. In addition to new investments in plants and equipment made by the firms, communities invested in the construction of the required physical and social infrastructure such as roads, bridges, utilities, railroads, courthouses, schools, parks, and hospitals. Private individuals invested their savings primarily in building and improving their homes and businesses, and fortunes were made in local development and in transit and utilities. Endowments were established for universities and cultural organizations and all types of new ventures were begun, including such things as hotels, expositions, theaters, professional sports, and golf courses.

NEW KNOWLEDGE AS CAPITAL

Investment also took some new forms that were to lay the foundations for a new type of power that would play a critical role in the formation of the industrial city and delay its developmental stage. As technology advanced, knowledge was accumulated. Industrial knowhow became a new form of capital. Investment in the production of knowledge can be referred to as human capital, to distinguish it from physical capital. The reason why this distinction is important is that physical capital is, like liquid capital, a commodity that can be controlled and moved quite separately from its owner, whereas human capital, which takes the form of knowledge or know-how, is strictly a human resource and does not exist outside the mind.

Human capital was created in many ways. Initially, human capital was created right on the production line by workers who acquired new skills and developed expertise and became highly skilled craftsmen. Human capital or knowledge was also formed by inventors and scientists working in laboratories and by engineers tackling new design and production problems. Management also accumulated a great deal of know-how as it learned how to manufacture, merchandise, and create markets for new products. Professionals also had to advance their knowledge; financiers were called on to help raise ever-increasing amounts of capital; lawyers were called upon to protect investments, to draw up contracts, and to litigate disputes that arose; accountants and management consultants were hired to monitor activities in order to find ways to cut costs and increase return on capital; labor unions developed organizational skills, developed pension and health programs for the workers; and the government was called upon to regulate the whole process.

All these activities contributed to the advancement of technology, and all this knowledge has to be passed on from one generation to the next. Thus we evidence the rise of apprenticeship programs, community colleges, universities, teaching hospitals, and corporate training centers to facilitate the passing on of knowledge. And as noted earlier, the passing on of a culture from one generation to the next is the principal role of the city in history. As industrial society becomes increasingly human-capital intensive, the industrial city plays an increasingly important role.

With time, as human capital or knowledge-intensive activities become more formalized, they will be easier to recognize, to understand, and to accept as being important. Some will become more formalized within the corporate structure, some externally in private firms, and others in organizations that now constitute the growing public and non-profit sectors of the advanced industrial economy.

Knowledge functions become increasingly differentiated within corporations as their operations and their knowledge base expands. New positions, new departments, new divisions, and even new companies are created as industrial corporations become larger, as their technology becomes more refined, and as their management capabilities become more sophisticated. Corporate offices, which may have begun as desks in corners of early factories have, in many cases, now become completely differentiated from manufacturing operations and are housed in separate and often very symbolic structures (the analogy to the statehouse, the temple, or bourse is appropriate). Research and development parks, engineering centers, tech centers, computer centers, training centers, regional offices, and divisional offices have all followed the same evolutionary pattern in their development. The design and construction of a new corporate headquarters or research facility is just another step in this long evolutionary process.

Outside of the corporate structure new types of advanced service firms evolved in response to the growing needs of corporations for particular kinds of expertise. Law firms, accounting firms, engineering firms, banks, insurers, advertising and public relations firms, printing and publishing, communications and travel specialists have all grown by serving the needs of nearby corporate clients. These professional, technical, and specialized business-service firms are able to provide expertise and advice that companies are unable to produce themselves in an efficient manner. These advanced service firms provide what economists call external economies. Several firms share the services of a group of specialists that no single firm could sustain alone (see Figure 3.1).

In time, some of these firms, which have grown by responding to the growing needs of locally based industrial firms, broaden their markets. They often begin by opening up offices in other cities in order to serve the expanded operations of their clients but then they begin to seek the business of industrial firms located in other cities. In

INDUSTRIAL CORPORATIONS*	SPECIALIZED TECHNICAL AND BUSINESS SERVICE FIRMS	PUBLIC AND NOT-FOR-PROFIT ORGANIZATIONS
*Includes purchasing and transportation		
Corporate headquarters **	Law	Federal agencies
Research and development	Engineering	State agencies
Regional offices	Accounting	Universities
Divisional offices	Finance	Musical arts
Computer centers	Advertising	
Training centers	Public relations	Hospitals and clinics
	Insurance	Cancer center
	Seminars and conventions	Professional associations
	Communications	Federal Reserve bank
	Airlines	Foundations
	Consultants	Museums
	Business information services	Consulates

FIGURE 3.1   Advanced Services Sector

fact, advance service firms usually have to expand and upgrade the services they provide in order to hold on to their corporate clients. Many engineering, advertising, legal, insurance, accounting, and financial firms have now evolved into international organizations.

These advanced service firms continue to expand, and together with the headquarters and related functions of industrial operations, now comprise the institutional or economic base on which the advanced industrial city is founded. These activities, referred to as the advanced service sector, represent the human capital base of industrial societies. The expertise and advice provided by members of these advanced service organizations enable corporations to operate in very complex, highly developed, competitive global markets.

Advanced services govern, that is, manage and advance technology; they are found at the leading edges of industrial society, they are engaged in problem solving, and advance the state of the art in the process. It should be noted that these services are purchased primarily by industrial corporations, and their cost is included in the final cost of the product manufactured. They should in no way be confused with consumer services, which are sold directly to the consumer in the community where the consumer resides.

It is the growth and increased concentration of advanced service activities both within and outside corporations that provide the basis for the future development of industrial cities. Even as industries mature, technology is continually being advanced and the knowledge content of goods increases. As industries mature and the role of advanced service activities increases, knowledge-intensive activities are becoming increasingly centralized because most industries are gradually restructured and eventually become dominated by a few major corporations. As smaller firms are taken over through mergers and acquisitions, their high-level management and technical functions are brought under the corporate umbrella and shifted to the city where the parent company is headquartered (see Figure 3.2 for an illustration of this process).

Industrial cities are developing in places where the headquarters of major industrial corporations and the operations of specialized advanced service firms are concentrated. These organizations constitute the institutional base on which the world industrial city will be founded. The function of industrial cities will be to govern the technology that corporations or organizations have developed and control. The analogy can be made to traditional cities; their function is to govern territory over which they have sovereignty or jurisdiction. Although power based on technology is new compared to power based on land or liquid capital, the scale and complexity of many corporations exceeds that of many nations. Both types of power require a large middle-class population to support the governance function. The governance function will continue to expand as the industrial know-how gained over the last hundred years of rapid industrialization is transferred to the developing nations. This is, of course, in addition to the continued advancement of technology and the growth of the knowledge content of goods and services produced worldwide. The deepening and spreading of knowledge is rapidly becoming the world's dominant activity.

Although the industrial city exists functionally it does not yet exist culturally; the prototype has yet to be built. What form this new type of city will eventually take is still an open question. Its institutional base—organizations that govern technology (in all its various forms)—has been established, but unlike traditional cities, their role in the world economy is still not understood, appreciated, or accepted in a cultural sense. Industrial cities are still viewed more as temporary encampments of manufacturing than as the capitals of technology.

The driving force behind the rapid growth of industrial activity has, until recently, been the growth of manufacturing operations, which was once synonomous with the growth of factories. All that is changing. Factories are being automated, computerized, and decentralized. But their knowledge base or brain remains and develops in the industrial city. The actual manufacturing facilities are being "deskilled" and relocated in "Greenfield" locations, usually in less-developed (lower cost) areas both in the United States and abroad. The factory, once the primary source of economic opportunity in industrial nations, is now moving to the newly industrializing nations. This is possible because the factor has become footloose, it no longer requires highly skilled workers and can be linked to its knowledge base by instantaneous worldwide communications. Consequently, manu-

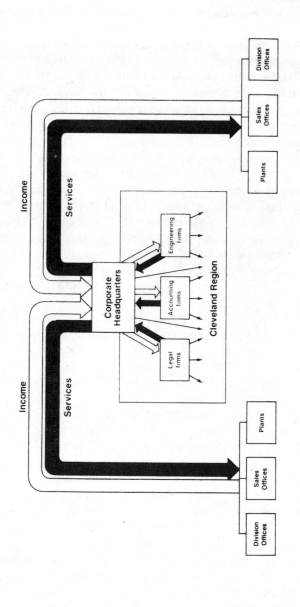

FIGURE 3.2 How Cleveland's Corporate Headquarters Generate Income for the Region Through Worldwide Operations . . . Supporting Jobs in "Advanced Services" and Local Sector

facturing operations are filtering down to less-developed areas where there is a better match between workers and skills required. In the future, we may even see certain types of manufacturing facilities in space where advantage can be taken of zero-gravity conditions. The point to be stressed is that as technology advances and as the knowledge content of goods increases, the actual manufacturing process can take place in areas remote from where the knowledge is produced.

THE PLACE OF FACTORIES IN URBAN HISTORY

Changes that have occurred in industrialized nations can best be viewed in the context of the gradual evolution of the factory. When most industries began, factories were small, often taking the form of cottage industries serving rural areas. Industrialization accelerated in urban areas and declined in rural areas with the building of railroads, the discovery of high grade ore deposits in the Upper Great Lakes region, reduction of shipping costs, and the invention of new steel-making processes. Large plants were constructed and costs declined; economies of scale were realized. The stage was set for the growth of large industrial complexes and the phenomenal growth of industrial activity. Although some manufacturing centers grew enormously, not all places or regions participated in the prosperity created by the ore boom and the advent of the steel era. While some canal towns were transformed into manufacturing and railroad centers, others languished, as did many stagecoach towns when the canals were first built.

Urbanized factories were originally located in close proximity to docks and railyards because water and rail transport were the most economical forms of transportation. Towns grew up organically around clusters of factories; neighborhoods grew within walking distance of the factory gates. Some communities were company towns, some had many firms but were dominated by one industry, others became diversified. The fortunes of these factory towns varied greatly depending on how locally based companies fared, on trends in their dominant industries, and general swings in business conditions. This era of rapid growth of the industrial workforce, of population, of housing, and of compact or high-density urban development close to factories, that in turn were close to ports and rail terminals, was, however, short-lived.

In hindsight it becomes clear that the early factory town or mill town was just the initial stage in the historical evolution of the industrial city. In the future, when industrial cities are better understood, the factory town will probably be viewed as an aberration in the historical evolution of cities. This is because they did not have one of the most critical attributes of cities: a dominant middle class. They were historical accidents; they began as manufacturing centers. The actual city-building process was to take several generations.

Initially, when manufacturing centers were populated primarily by factory workers, power was concentrated in the hands of a few wealthy owners. Most of the workers were recent immigrants either from foreign countries or from rural areas. They were usually unskilled, with little or no formal education or savings. The neighborhoods were built according to the needs and resources of the working class and they soon became obsolete as workers, who upgraded skills and income and gained middle-class status, moved to middle-class residential areas in the suburbs developed to meet their rising expectations. New transportation technology, namely the streetcar, electricity, the telephone, and later the automobile and television really opened up the suburbs. Subregional commercial and retail centers sprung up at the major intersections of major highways, drawing considerable activity out from the central city. The role of the city as a place to live, shop, and play was rapidly undermined.

This thumbnail sketch of city development in industrial areas over the past century would be incomplete if the context in which this development occurred is not considered. Some very significant social and political developments were occurring at the same time. Attention needs to be drawn to contextual issues because the same forces are not likely to be present in the next century. First, the Industrial Revolution had a differential impact on cities in different regions and as a result resources shifted from one region to another based on locational advantages. In the future, as industry becomes footloose, and as the skill requirements of workers in these factories are reduced, we will find fewer and fewer factory jobs in cities, even in the older industrial regions. Just as industrial cities benefited from the buildup of manufacturing jobs during the early stages of industrialization, they now suffer most from the decline of production jobs. This decline will continue as industries mature and as they are rationalized on a world basis. Almost all cities that have large manufacturing sectors are currently seeing a reduction in their production workforce. The difficulties that this erosion of production jobs poses to the community depends on how dominant the manufacturing sector is and on whether nonproduction activities in manufacturing and in advanced services can expand sufficiently to offset declines in production activity.

THE TRANSITION TO ADVANCED INDUSTRIAL CITIES

It should be noted that a decline in production workers does not necessarily result in a decline in metropolitan income or total employment. The wealth created or valued added by production facilities may actually increase because it is usually the unskilled and the lower-paying production jobs that are lost. Production activities that remain usually involve high levels of skill and are upgraded as technology advances. As long as the value of exports increases, income flowing into the region will continue to increase in real terms, even though the number of workers required to generate this income may decline. The total number of jobs in the community increases because of the income and jobs created by the multiplier effect of increased earnings in the export sector. As industrial activity is upgraded, economic opportunities for the less-skilled and for new entrants thus shift from factory to office settings and to local services such as government, retailing, construction, repair, food services, health, and a wide range of locally oriented services.

The shift to a service economy is a national and international phenomenon. Although the growth of services is most evident in cities that do not have large manufacturing sectors, services are also growing in manufacturing centers that are in decline. However, the gains made in service industries are offset by the decline in routinized production jobs. Unless the transition is understood, the nature of development in industrial cities will not be appreciated and the growth of the service cities, which are primarily but not exclusively in the sunbelt, will be over-emphasized. It is very important that the nature of the development and transition of cities in the older manufacturing region is understood by the general public because, should these cities be viewed as being in decline, the success of their difficult transition would be placed in jeopardy.

Industrial cities are easily misread. If one thinks of them in terms of plant and equipment, all one sees is abandoned plants and neighborhoods in the center of the city, and thus their image of inevitable decline is confirmed. If one thinks of them as having a strong middle class living in the suburban settings, commuting to office jobs in the CBD, and enjoying well-established and high-quality cultural and social support systems, then one's image of the emerging world city is reinforced. The image of the industrial city is extremely important especially when firms are recruiting talent for advanced service activities in competitive world markets. Image also plays a role in the decision making of local residents, organizations, and communities as they decide where to go to college, begin a career, form or move their family, or begin or expand a business and invest in improving community facilities. A negative image can become self-fulfilling and a major impediment to development. Viewed in perspective, the worst is over. Most of the production jobs that will be lost have already made the most difficult adjustments. The challenge they now face is one that all cities have faced in history: to rebuild. This requires a clear vision of how their role has changed so that they can be redesigned in ways that are supportive of their expanding functions.

FORCES OF URBAN DEVELOPMENT

City development has been very sporadic in the twentieth century; industrial cities flourished as sunbelt cities languished during the first quarter-century, and all city development came to a standstill during the Great Depression. Development was further distorted during World War II and the post-war and the Cold War era as the old plant was called on first to provide military supplies and later to supply capital goods for rebuilding Europe and Japan under the Marshall Plan. It then rebounded at an accelerated pace for roughly a quarter-century of increasing affluence before leveling off in the last decade. Structural changes are no longer camouflaged by rising levels of employment and real income everywhere; difficult adjustments are occurring.

Underlying these trends were five unusual phenomena that need to be highlighted. One phenomenon was the baby boom resulting from deferral of family formation during a quarter century of depression and world war. Another one was sociological: the realization of the widely shared suburban dream, and the concommitant problems of decline in older inner-city neighborhoods. The third was political, and concerned the growing role of government in an increasingly pluralistic economy, and a polarization of the public and private sectors. The fourth was economic and brought steadily rising real incomes resulting from rising productivity, expanding participation in the labor force, low interest rates, and high levels of confidence and investment in the economy. The last phenomenon was a technological one resulting from scientific advances over a wide front, especially during the Cold War, the Space War, and particularly in defense-related fields such as weaponry, aircraft, electronics, computers, and communications.

When we contemplate the forces that will influence city development in the future, it is natural to project on past trends; however, these could be very misleading. There will not be further migration from rural to urban areas, and it is unlikely that the rate of population increase will return to its high post-war levels; therefore, instead of cities spreading out into the suburbs and merging into gigantic megalopolises, as was envisioned mid-century, we will see development returning to the cores of cities, taking qualitative rather than quantitative dimensions. In short, the shift of labor from field to factory is over. The rationalization of production activities on a worldwide basis is well underway, and the shift of work from the production line to technical and managerial centers is gaining momentum. As the knowledge content of goods and services increases, the knowledge worker, found primarily in office-type settings, will increasingly dominate the national and the city economy.

Changes in the industrial structure, and expansion of knowledge-related functions performed in cities, will continue to produce a change in the nature of work and in the place of work. The expansion of office-type work space (interpersonal-communication and information-intensive activities) will continue, even if not in high-rise office buildings. The upgrading of the work force, both in white-collar occupations and in blue-collar occupations, will place new demands on human resources, cultural support systems, and on community development programs.

The traditional industrial development strategies that emphasize the recruiting of plants, plant expansion, and retention will not produce sufficient employment opportunities in the future. There will still be production-type jobs in cities, but they will be primarily be positions for highly skilled craftsman, and these jobs will require an extensive apprenticeship program and high academic achievement. Skilled craft jobs can best be regarded as a knowledge-intensive activity; they will continue to be financially rewarding and should be viewed as supporting middle-class households, not working-class households. It is interesting to note that the basic wage of steel and auto workers in 1982 is about $44,000 (including benefits and not including overtime). The image of the blue-collar worker as "Johnny six-pack" is outmoded. Blue-collar work will continue to gain in status as the routinized, repetitive jobs are replaced by robots. Unskilled jobs in manufacturing will continue to exist but they will not be of a permanent nature. They will appear principally in new firms that are incubating and in manufacturing firms that serve a regional market.

The real success of any city in the future, as in the past, will depend on its ability to pass on its knowledge base from one generation to another, to retain its young once they have been educated (the major investment any community makes), to recruit talent to apply and advance the wide range of technology and know-how that is required by both industrial and nonindustrial organizations, and also on the city's ability to retain the retired, who have spent their working lives in the city. This suggests that the development strategy for cities in the future will have to be oriented toward rebuilding a city that is amenity-rich, that is, attractive to the middle class. Although this may seem like the obvious direction for industrial cities to take, it would require a major shift in values, behavior, and policies that have evolved and been practiced over the last century. Industrial cities have been product, not people oriented; making these places livable and competitive with world-class cities is the challenge of the century.

NEW FORMS OF URBAN POLICY

Urban policy has, during the expansionary phase of manufacturing, been problem-oriented and of a reactionary nature. As a result most of the local policies have been geared to keeping costs and taxes low as a way of promoting the expansion of production jobs. Federal programs have been oriented toward their problems. In recent times the problems have been over-publicized by cities as a way of making a case for more federal aid. This will have to change if cities are to become opportunity-oriented and aim to realize their potential. What is needed is a development strategy that identifies opportunities inherent in the development of each particular industrial city and its surrounding region. This change in orientation, from city decline to city rebuilding, from problem seeking to opportunity identification, from resisting change to anticipating and accomodating change, from short-sighted self-interest and special interest to long-term community and regional interest, from retrenchment and managing decline to upgrading institutions and the environment, from reacting to external forces to taking local initiatives and risk, will require a profound change in attitudes and in civic and political behavior. Clearly the critical attribute of success will be leadership, especially the ability to envision the future of world cities, and to formulate strategies for their design and development. Leadership not in the form of powerful statesmen, as was the norm in the past, but in the form of an enlightened citizenry.

The creation of cosmopolitan sectors in industrial cities, and thus the solving of most urban problems, hinges upon the return of middle-class residents to the city. This is unlikely unless cities formulate a long-term comprehensive urban development strategy directed toward these ends. Residential redevelopment is unlikely to occur without a long-term strategy that has the support of all political factions in the region, both within the central city and without. The strategy will have to reverse the long-established urban growth patterns whereby resources are pulled out of the city into peripheral areas where higher quality environments can be created and maintained. Market forces that favor the return to the city, such as rising costs of commuting (especially time costs), of land, of construction, and of service provision in low-density areas, are not sufficient to reverse these patterns. There are many who place their hopes on what may best be viewed as a spontaneous remission; this has occurred in a few special, highly publicized locations such as Society Hill in Philadelphia, Back Bay in Beacon Hill in Boston, Greenwich Village and Brooklyn Heights in New York, and Georgetown in Washington, D.C. Neighborhoods originally built for middle-class residents, well-planned and built to high standards, can be restored, but areas that were unplanned and built to house new immigrant workers have long become obsolete, and disinvestment in these areas has made them unsuitable for revitalization by market forces alone. It should be noted that all great cities have rebuilt more than once; each time new values are articulated and the city becomes more cosmopolitan in outlook.

THE ROLE OF THE MIDDLE CLASS

The return of the middle class will contribute to the city's redevelopment in several ways. First, the city's tax duplicate will be expanded and thus provide the city with a way of resolving its fiscal problems and enable the city to finance improvements in infrastructure (roads, bridges, utilities, schools, parks, etc.) that are so badly needed in the older industrial cities. Second, job opportunities for low-income city residents will be expanded. Once the middle class is living in the city and spending its leisure time and earnings in the city, jobs will be brought back from the suburb to the city in close proximity to members of lower-income households. Whereas these households often depend on such opportunities for their livelihood, suburban households use these jobs to supplement the earnings of the primary wage earner, or simply as an entry point into the opportunity matrix. Third, the presence of the middle class is very important in terms of leadership resources needed by a variety of institutions, particularly large inner-city school systems, career guidance and manpower planning and training agencies, and economic and community development agencies.

Many of the institutions that were established in the early phases of the cities' development are still aligned to sectors now in decline and thus do not prepare their constituents for opportunities being created in the expanding sectors. Some organizations have even become dysfunctional and suffer from bureaucratic rigor mortis. Instead of providing institutional linkages between declining neighborhoods and employment opportunities, they have become obstacles that their constituents have to overcome in order to gain access to new jobs in expanding sectors. In contrast, suburban-based institutions are closely tied and very responsive to changes that are occurring in the labor markets, and thus their graduates have better access to new opportunities even when they are located in the central city than do graduates of inner-city organizations that are in closer proximity. Moreover, labor markets are becoming increasingly competitive as female participation rates increase, and as the percentage of households having more than one wage earner increases. The added flexibility that these developments now give to labor markets makes it even more imperative that institutions serving inner-city residents become more responsive to changes in the labor market. The return of middle class residents to the city and their active participation in the determining of policies by city agencies is absolutely necessary if the needs of low-income citizens are going to be addressed. The old view that the disadvantaged need only occasional training needs serious reconsideration. The old stereotypes of blue-collar workers (Johnny-six-pack) do not do justice to today's industrial workforce. Outmoded images of industrial workers are gradually giving way to a more enlightened view of industrial workers as constantly engaged in advancing the technology they employ rather than those who persist in resisting change. These changes in the labor market should be viewed in a positive fashion. An effort should be made to accomodate the structural and occupational changes that occur. All members of the workforce will need to be able to upgrade their skills through continuing education. The expansion of local community colleges, proprietary schools, universities, and educational programs offered by a variety of cultural institutions will become increasingly important to those who desire to acquire new skills and develop their interests. New institutions are being established in response to the changing nature of work and to changes in attitudes and lifestyles, and this augurs well for the future of the industrial city.

THE INTENTIONAL CITY

As the industrial city is redesigned and rebuilt it will become a more intentional city. The industrial city could not have been planned earlier because it could not have been predicted that Manchester,

Birmingham, Detroit, Pittsburgh, or Cleveland would become great cities. But when they are rebuilt to meet contemporary needs of their citizens, the old priorities of keeping costs and taxes down (which is necessary if a city is trying to attract factories in order to create jobs for low-skilled workers) will be replaced, and high priority will be given to creating the amenity-rich environments needed in order to develop, retain, and recruit talent required by knowledge-intensive activities. The industrial city will in the twenty-first century challenge and perhaps even surpass traditional cities as desirable places to live. The quality of services provided by private, public, and not-for-profit sectors and the quality of the environment in general will become increasingly important as new types of world cities evolve and as the ranks of the knowledge workers expand. High-quality support services and a wide range of educational, social, and cultural services will be essential to the maintenance of high productivity in knowledge-intensive activities. Continuing education programs and high-quality medical services will be the norm. In short, industrial cities will become more people-oriented, and their success will depend on how attractive their environments are in relation to other cities that are competing for knowledge workers. (This does not mean that taxes will be low, but rather that services will be of high quality and efficiently managed so that they are supported by local citizens and remain competitive to those offered by other cities vying for talent.)

Whether an advanced industrial city can identify and act upon opportunities created by the emerging international order will depend to a great degree on how economic development is perceived, on how necessary changes are viewed politically, and on whether the city is psychologically prepared to take on a role of managing and advancing technology even while traditional manufacturing activities are declining. Cities that make a commitment to developing the capacity to anticipate and accomodate changes brought about by industrial transition will be able to minimize its adverse effects and reduce the fears and anxieties that will inevitably accompany these changes.

Cities that understand the nature of the development processes and the forces underlying their development, that have nurtured a sense of pride in their past accomplishments and actively seek to educate their citizens both about their city's history and about the challenges that they face in the future, will be well poised for the next century. Each city needs to evaluate its past and learn from its experience to prepare for the future. The future of the city will be created by the concerted actions of citizens as they reinvest the wealth they have created in the past in ways that will insure their city's future. It is not a particular vision of a city's future that is important, what is important is that the citizens of each city become aware of the role that their city now plays or desires to play in the future and that this responsibility for the future is accepted by the citizenry at large.

REFERENCES

GINZBERG, E. [ed] (1973) New York is Very Much Alive. New York: McGraw Hill.
GREENFIELD, H. I. (1966) Manpower and the Growth of Producer Services. New York: Columbia University Press.
KNIGHT, R. V. (1977) The Cleveland Economy in Transition, the Role of Corporate Headquarters. Regional Development Program, College of Urban Affairs, Cleveland State University.
——— (1973) Employment Expansion and Metropolitan Trade. New York: Praeger.
NATHAN, R. and C. ADAMS (1976) Understanding Central City Hardship. Political Science Quarterly 91 (Spring): 51.
The Northern Ohio Urban System Research Project (1973) A Concept Plan for Future Development, Final Report, Report no. 20, Athens, Greece: Doxiadis Associates International.
PETERSON, G. E. (1978) The Future of Cleveland's Capital Plant. The Urban Institute.
PIRENNE, H. (1970) Medieval Cities, Their Origins and the Revival of Trade. Princeton, NJ: Princeton University Press.
STANBACK, T. M., Jr. and R. V. KNIGHT (1976) Suburbanization and the City. Montclair, NJ: Allenheld, Osmun.
——— (1970) The Metropolitan Economy: The Process of Employment Expansion. New York: Columbia University Press.

4

Seven Scenarios of Urban Change

ARTHUR SHOSTAK

☐ IN APPROACHING THE ISSUE of urban change, one can either be more of an optimist or a pessimist. Optimism in the future of at least some cities is warranted in the minds of some observers by seven different forces including

(1)   the emergence of a new urban leadership that views the tasks of government as technological, managerial, and highly manageable;

(2)   an increase in energy and fuel costs discouraging living in suburbia and driving to city jobs;

(3)   the continued slowdown in any significant migration of underprivileged blacks and whites off the land and into northern cities;

(4)   the steady acculturation of all previous newcomers to the cities, as exemplified in the stabilization of public school test scores and the reduction in family size of low-income urbanites;

(5)   the steady improvement in America's race relations, as demonstrated by Gallup Poll data and by the recent absence of violent racial confrontations;

(6)   the seeming gains of teenage maturity, as evidenced by the sharp decrease in teen gang killings and by the ability of thousands to coexist peacefully at major entertainment events;

AUTHOR'S NOTE: *This chapter was adapted from report prepared for the Philadelphia: Past, Present, and Future project conducted by the Center for Philadelphia Studies, 4025 Chesnut Street, Philadelphia, PA 19104. The report, "Long Range Future: Seven Scenarios for Philadelphia's Next 25 Years," is available for $4 from the center, as are several other similar task force reports on various aspects of Philadelphia urban life (including "Culture and Recreation," "Energy," "Transportation," "The Peace of the City," "The Purse of the City," etc.).*

(7)   the persistence of ethnic and racial pride in neighborhood ambience and well-being, as exemplified in the large and growing
number of block associations, neighborhood organizations, and
so forth.

A more pessimistic view is that the city of the future will never resemble the brightest dreams of the city held in the past. This view lacks faith in traditional planning if it remains unaccompanied by a radically new political and social vision. Without a dramatically modernized vision and the will to enact it, American cities could, in 25 years, become like the fortress city so graphically described in the Kerner Commission Report (1969), a city of incessant strife; of abandoned, deteriorating, dangerous neighborhoods surrounding a highly fortified, high-income central business and residential district; a city divided drastically into mutually fearful and warring Haves and Have-Nots.

An utterly frank dialogue between urban optimists and urban pessimists might agree that the urban future isn't what it used to be, a recognition that would help bridge the different perspectives of urban change present in both Sunbelt and Snowbelt cities. Even in the glamorous new cities of Houston, Atlanta, and Denver, concerns are steadily growing over crime rates, the sterility of the downtown, and the influx of unskilled workers.

In spite of short-term uncertainty, then, about the prospects for American cities, it is possible to project several emerging patterns of future change or alternative urban scenarios that cities can choose from. These are shown in Table 4.1, and each highlights an underlying arch issue, central problem, or key opportunity faced by most major cities today.

The first of these scenarios—Conflict City—represents the current state of affairs for many American cities, especially those of the midwest and northeast. The other scenarios are all foreshadowed by existing realities and emerging possibilities: Wired City, for example, represents the extension of high-tech modernity, while Neighborhood City is an expression of the new American urban ingenuity represented by the "urban homesteading" movement. Conservation City is a further expression of the forces rampant in the response of snowbelt cities to the rising costs of energy. International City is a scenario already manifest where the headquarters of multinational cor- porations reside, and it reflects a new business urbanity. Regional City, in turn, is less common, its rarity explained by its role as an expression of new patterns of political power-sharing and cultural-

Table 4.1

	Scenarios	Underlying Issues
I.	Conflict City	What might cities be like if we do as little as possible to change things?
II.	Wired City	What are the rewards and risks inherent in mind boggling communication breakthroughs?
III.	Neighborhood City	What fresh role might local communities play in meeting tomorrow's urban needs?
IV.	Conservation City	What can we do to turn energy shortfall to our collective advantage?
V.	International City	What can we do to assure more gains here from an emerging global economy?
VI.	Regional City	What sorts of emerging urban problems are best met by a regional response?
VII.	Leisure City	What might urban life resemble when free time and recreation goals increase in our lives?

leisure interaction. Finally, Leisure City represents a precursor of a new form of "right brain" playfulness engendered by the time released from work routines by new technologies, new forms of work organization, the prolongation of adolescence, and the extension of longevity among the growing number of people over 65.

Each of these scenarios will be discussed in turn. Other scenarios, such as Senior Citizen City, Human Services City, Abandoned City, or Learning City could also be suggested, of course, but the seven chosen for discussion seem the most representative of events and trends in the next quarter-century. As shown in Table 4.2, each can already be associated with contemporary conditions in a number of places.

While these embryonic examples may only hint at the 25 years of future development envisioned in the various scenarios, each of these prototypes can suggest different policy lessons of value to concerned urbanites in the here-and-now.

Before each scenario is discussed in more detail, it might be useful to review the nature of scenarios and their use, as this intellectual tool puts unique demands on users. Accordingly, three guidelines below may help make a stronger reader of each of us.

To begin with, scenarios ask us to suspend a natural faculty we have for inferring the likely future from the present. Extrapolating in

Table 4.2

	Scenarios	1981 Counterparts
I.	Conflict City	Detroit; Youngstown; Camden; any large Frostbelt city in a "steady state" stage
II.	Wired City	Manhattan in New York City; London's Central Business District; Washington, D.C.
III.	Neighborhood City	Santa Barbara, California; Seattle, Washington; Boulder, Colorado
IV.	Conservation City	Habitat (Montreal); Hong Kong; Tokyo; Manhattan in New York City
V.	International City	Paris; Toronto; Manhattan in New York City
VI.	Regional City	Greater Miami; Greater Toledo; Greater Toronto; Minneapolis-St. Paul
VII.	Leisure City	Reno, Nevada; Palm Beach, California; Orlando, Florida; Anaheim, California

a straightforward fashion should be replaced by a more creative and surprise-open attitude:

> Most people—including many futurists—conceive of tomorrow as a mere extension of today, forgetting that trends, no matter how seemingly powerful, do not merely continue in a linear fashion. They reverse direction. They stop and start. Because something is happening now, or has been happening for three hundred years, is no guarantee that it will continue [Toffler, 1980].

Each of the seven scenarios, in short, requires a willing suspension of disbelief, an appreciation of the notion that the future is not fully previewed by the present (which, in any case, is itself "exploding with paradox").

Second, the speculative nature of various scenarios often precludes their containing field reports of research findings. Typical is the situation where two-way TV is concerned:

> The social effects of viewer-to-source response systems on their communities are a matter of speculation. There have been no studies as of Summer, 1980, and the wired communities themselves are too recent to manifest easily observable changes.... Moreover since there are as yet only a few experimental wired communities and these communities know themselves to be experimental, and, perhaps, think themselves "special" or "temporary," the applicability of this experimental data to regularized programming is questionable [Pergler, 1980].

Each of the seven scenarios offers far less in hard evidence, pro or con, than we might prefer; and while systematic and evidence-based material is more familiar and always instructive, "in the end we must embrace—not dismiss—paradox and contradiction, hunch, imagination, a daring (though tentative) synthesis" (Toffler, 1980).

Finally, in the spirit of necessary intellectual risk taking and in full recognition of critical gaps in the research record, scenarios should obligate us to tentatively take sides in major and open controversies. Typical are arguments pro and con on the significance of computers in the urban future, a question that goes to the very heart of the Wired City scenario. Skeptics contend that computerization is less profound in its effects on people's lives than the changes earlier wrought by our adoption of such technologies as railroads, elevators, or bulldozers. Given, however, the rapid rate of market penetration, the daily expansion in rewarding software, the downward spiral of computation costs, and the escalating public expectations of this new technology, most of us will ultimately side with those who forecast more, rather than a modest or even a minimal impact by computers on urban life tomorrow.

These three guidelines: (1) the advisability of keeping an open mind, (2) making the most of data-sparse speculation; and (3) accepting the (tentative) taking of sides, should help the reader take from the scenarios a rich sense of urban future diversity. There is no single urban future over the horizon, but many from which we must choose...lest we aimlessly and dangerously drift.

CONFLICT CITY

As the Conflict City scenario represents the current status of the post-affluent transition in many of the older American cities, it has been a familiar reality to many of us for the last two decades and will therefore not be discussed in any great detail. Conflict City was foreseen by the Kerner Report (1969), which warned against allowing the current governmental paralysis to continue. Its unspoken principle in guiding public expenditure decisions is triage, a policy that permits certain devastated areas of the city to go without public services and succumb to a state of human anarchy.

The policy of Conflict City remains seriously divided. While the labor force competes for a shrinking pool of employment opportunities, neighborhoods lose the best of their young people, and fear and distrust render the normal disorder of the city almost unbearable. Unfortunately, efforts to reverse trends contributing to this scenario are retarded by backward-glancing commitments to dreams of "how things used to be" and outmoded, aristocratic visions of "what cities are supposed to be."

WIRED CITY: AN ELECTRONIC COMMUNITY

For some observers the most intriguing and perhaps the most compelling urban scenario has rapidly become that of the Wired City. New technologies of the communications revolution may, by the start of the twenty-first century, contribute to a fundamental urban transformation, and many urban activists want to directly guide the related changes.

Direct communication by telephone in the late 1800s was that century's only supplement to earlier reliance on indirect transmission of messages by letter, pamphlet, broadside, newspaper, magazine, telegraph, and photograph. Early in the twentieth century, however, cities began to adapt to a variety of new mass-communication modes—first movies, then radio, then "talkies," and, later, what has been described as the world's most powerful communications medium: television (Martin: 1978).

Looking over the next 5 to 25 years, the newly arrived information age, with its growing interconnection between computers and telecommunications, has the "possibility of transforming urban society in the way that railroads and electricity did in the nineteenth century.... The 'computerization society' will shape, allow, facilitate, determine...an extraordinary transformation, perhaps even greater in its impact than the industrial revolution of the previous century" (Simn: 1980). Indeed, Toffler ranks computers "among the most amazing and unsettling of human achievements, for they enhance our mind-power as the Industrial Revolution enhanced our musclepower, and we do not know where our minds will ultimately lead us" (1980).

A Philadelphia task force undertook a thorough study of the significance of the information age in Philadelphia—esepcially in terms of the policy questions raised by its advent. Attention was paid to separate, though related impacts of mass media broadcast hardware such as radio and TV with their mass market and mass culture orientations, and the newer telecommunications "narrowcast" hardware such as electronic conference equipment and video-text receivers, which permit the near-instant transmission of messages (information) from one point to another specific point or group of points, rather than to an undifferentiated mass, as in the past. The study group resea ched the newly engineered ability of computers to process verbal, as contrasted with numerical information. This text processing ability has a growth potential vastly greater than more traditional "number crunching":

> Soon there will be computers in our automobiles, our typewriters and sewing machines, our air conditioners and television sets and stereos. We will wear them on our wrists. They will be ubiquitous, and we will use them without thinking about them. As a result, we will be different people, doing different things [Russel, 1978].

Fascinating in this connection are the claims made by commentators utterly persuaded of the revolutionary impact of an information age:

> The revolutionaries are the people who have learned how to exploit the new micro-technology of computer-linked communications to the fullest: manufacturers who own no workshops, bankers who handle no cash, retailers who advertise goods "not available in any store," and businessmen and women who "walk" to work rather than ride. What is now unfolding is the most massive restructuring of the city's economy since the Industrial Revolution [Smith, 1981].

Similarly, computer fans find it easy to imagine a typical day in a city restructured by telecommunications:

> It's 6 P.M. You drive into your driveway. The bordering lights and front door lamp light-up automatically. At the front door, your voice command releases the special burglar-proof electric lock.
>
> Inside, you select the evening menu from the nutritionally balanced suggestions shown on your home computer video screen. In the den, the children settle down to study their math and French lessons at another terminal of the household computer.

Critics, of course, wonder aloud if this really amounts to the good life, as too many it appears cold, impersonal, mechanical, and manipulative.

From the very outset of their research the Philadelphia study group was offered contradictory emotional claims and complaints alike from enthusiasts and critics. Humanist supporters, for example, saw in the information age an "alternative to industrial civilizations, and the possibility of a technologically advanced but sustainable society"

(Smith, 1981). Technology proponents hailed our newfound ability to actually do essentially anything we could conceive of doing electronically:

> Some ideas which come easily to mind are remote catalogs and ordering from the home; go-anywhere-telephones; fingertip information on nearly any topic; machines which speak and listen; three dimensional life-size television; complete automation of appliances.

> But these are trivial examples, constrained by our present way of thinking and doing things; having electronics a million times more capable than that of today opens possibilities that we can now only begin to conceive [Gordon, 1981].

The question of whether or not the Wired City scenario is really for everyone is where many serious critics begin, even as some go on to raise vexing questions about a Wired City's dehumanizing possibilities:

> The combination of powerful home computers and stunningly effective three-dimensional video might provide totally credible psuedo-challenges, and by doing so completely blunt the edge of human curiosity and dynamism [Evans, 1979].

Others have serious qualms about totalitarian possibilities:

> There's going to be an awful lot of tracking of people's tastes and habits, and we have to be very careful that human rights are protected. If somebody in power wanted to establish a witch hunt at a particular time...the availability of information through cable subscriptions is far more serious than anybody having access to your Diners Club card or your bank account. And that is something we have to watch [Winfrey, 1981].

Above all, proponents and critics alike agree that many complexities remain to mar the entire scene, and that atavistic politics, single-issue lobbying, monopolistic sloth, regulatory ignorance, or vested interests could allow the least/best scenario to unfold...least/best from nearly everyone's point of view.

Given the potential rewards and risks of this Wired City scenario, it is useful to speculate, even now, that cities could profit from ten major urban aspects of tomorrow's telecommunications put in their best possible light:

(1)   Improved education, as provided by television hookups complete with home computer reinforcement. Programs could bring outstanding educators at a learner's convenience, day or night, seven days a week.

(2)   Dynamic politics, as promoted by two-way television that would enable viewers to question candidates, as also in electronic town meetings and in electronic referendums that provide rapid measures of public opinion.

(3)   Home-centered consumerism, as supported by the reservation of certain cable TV stations for selling goods and services, and by the establishment of electronic funds transfer systems (EFTS).

(4)   Reduced travel needs, as provided by the transfer home of certain types of computer-based work, certain types of purchasing options, and certain forms of mass media (first-run movies, plays, etc.).

(5)   Improved communication, as facilitated by two-way television, TV-aided phone systems, electronic newspapers, and computer conferencing.

(6)   Leisure plentitude, as offered by possibly more TV stations, videodisks, and video-recording options than individuals can ever possibly employ.

(7)   Twenty-four-hour work day options, as facilitated by the continuous-operation dynamics of telecommunications hardware.

(8)   Improved health care, as bolstered by electronic monitoring of vital health signs from one's living room, the linkage of hospitals everywhere to the world's specialists and specialized computers, and the provisions of home health care education via cable TV offerings.

(9)   Public safety and household payment gains, as provided by telecommunication systems for fire, home security, or medical alerts, and the remote sensing of home utility meters.

(10)   Personal empowerment, as facilitated by the provision of rapid and relevant feedback, information, and guidance.

Even a list as short and tersely put as this helps explain the enthusiasm of pro-Wired City urbanites.

To help minimize hazards that can be anticipated now and, likewise, to help amplify the scenarios' potential gains, constructive critics of telecommunications urge urban decision makers to get out ahead by

(1)   promoting diversity in information outlays, as in requiring cable-TV franchise winners to set aside valuable time for the use of community groups, public interest organizations, and other noncommercial, nonprofit-oriented operations (Rutter, 1981);

(2)   promoting parity in social class access, as in the establishment of a municipal rental service to help insure that have-not urbanites have access to home computer systems and other aids to human capital development (Friedman and Wildavsky, 1979);

(3)   promoting variety in systehs ownership, as in the mandating of cable-TV station allocation to minority residents (race,ethnicity, sex, handicap, etc.), either in the fact of ownership or in the selling of shares (Serafini and Michael, 1980);

(4)   promoting social impact assessments of the new communications aids, through contracting for city-based evaluation studies;

(5)   promoting local dialogue about Wired City national policy issues; and

(6)   requiring the implementation of retaining and retraining incentives through nw legislation requiring companies that switch to telecommunicatios hardware to offer upgrading employment options to the affected employees.

Especially as an information-based economy may prove more of a labor-releasing than a labor-utilizing one, early attention to potential labor force chaos cannot be urged strongly enough: "Electronic switching systems, bubble memories, and microprocessors now enable New York Telephone to serve 30% more callers than in 1965, with 15,000 fewer employees.... The common man has virtually no future in the brave new city. White-collar, high-technology jobs are no help to the unskilled poor" (Smith, 1981).

The Wired City scenario, in sum, suggests that American cities face a communications and information revolution whose impact is likely to rival that of printing 500 years ago. Any ensuing telecommunications upheaval clearly offers a remarkable range of gains:

Better education, better news media, improvements in the political process, better forms of human communication, better entertainment, better medical resources, less pollution, less human drudgery, less use of petroleum, more efficient industry, and a better informed society with a rich texture of information sources [Martin, 1978].

If cities are soon to learn how to profit from new uses of information (and its components—facts, opinion, persuasion, decisions), a Wired

City scenario must soon receive citizen review and experimental employ throughout the United States. Individual cities, in conjunction with universities, foundations, and corporations, should begin to organize their response to these extraordinary opportunities, especially to their anticipated and less-readily anticipated consequences.

NEIGHBORHOOD CITY: HUMAN SCALE

A considerably different scenario is preferred by those who are persuaded that, after the family, the neighborhood remains the basic building block of American society. This Neighborhood City scenario calls for "local self-reliance, and a return to the human scale of society, where decision making can be done through face-to-face contact, and where production takes place within walking distance" (Morris and Hess, 1975). In this scenario "neighborhoods must,... while struggling to change outside intervention in local affairs, absentee ownership, cultural dependence,... also strive to develop, for themselves, new institutions, new relationships, new mechanisms for providing basic goods and services" (p. 15).

At the heart of this scenario is a distinct vision of a humanistic urban community and new forms of appropriate technology including windmill units, solar greenhouses, methane generation from garbage, recycling units, community food processing, urban gardens, fish farms, poultry yards, jitney services, and so forth.

In some cities various neighborhoods would individually and jointly experiment with a new mix of technologies—some might specialize in food raising, others in handicrafts, others, in the provision of human services—even while all used appropriate technologies to secure more decentralized autonomy than ever true of such urban areas.

If, over the next 25 years, we experience any combination of the major setbacks threatened by energy, ecology, and food supply vulnerabilities, certain cities may have little choice but to adopt the Neighborhood scenario. Some futurists, accordingly, expect a quarter century (1981-2006) in which society will be obliged to make frugal and fruitful use of all its resources: recycle used material, reduce the expensive transit of goods and workers, maximize conservation measures, employ renewable fuel substitutions for steadily disappearing fossil fuels, explore the substitution of hand labor for certain energy-greedy mechanical processes, and so on) (Berg and Tukel, 1980). In this quarter-century

the timeless virtues of loyalty, cooperation, and selflessness—all thinly observed now, will once again be functional, as will the simple pleasures of family and friends, the knowledge of a trade, and the comforts of a well known environment [Johnson, 1979].

In such a tradition-shattering and energy-short context, urban dwellers may come to regard austerity and frugality as new moralities, new aids to right conduct.

Leadership would be expected to come from a constituency created earlier in the 1970s:

> Day-care centers, food cooperatives, environmental organizations, minority caucuses, senior groups, consumer action networks, handicapped advocacy organizations, and a multitude of block- or neighborhood-based civic action groups...have a natural inclination toward decentralized applications, and many have been active in advocacy or development [Corbett, 1981].

Together this new coalition—oldtimers and newcomers alike—would tackle problems of food and fuel shortages, and would recycle possibilities and human services needs with fresh ideas and grass-roots venture.

Strategic here is the development of unusual potential for a neighborhood-based economy:

> New disciples of local control point out with increasing evidence that modern technology itself is decentralizing in its impact and that now, perhaps for the first time in history, we can begin to speak of efficient productive facilities on the local level [Morris and Hess, 1975].

Finding ways to gain leverage over local economic forces, however, and to establish community financial institution (credit union, community development corporation, etc.), along with decision-making neighborhood manufacturing, will require much imaginative study and bold action.

Critics downgrade this entire scenario on several counts, and a healthy controversy persists on all fronts. Typical is the argument that any neighborhood uplift may encourage large-scale gentrification, a process that has well-off, generally Caucasian "invaders" dispossessing generally nonwhite have-not urbanites who are sorely victimized in the process (Laska and Spain, 1980). Proponents insist that "incumbent upgrading" is a viable alternative to unwelcomed gentrification or "suburbanization" of city neighborhoods, and they champion upgrading (repairs and reinvestment) without any unreasonable turnover in neighborhood occupants.

Critics make much of their contention that modern architecture opposes the neighborhood scenario, as it reflects bureaucratic, centralizing values. Proponents, however, dwell on their hope for a postmodern architecture, an emerging force that embodies and recapitulates local history, respects local context, resurrects the public realm (the assembly area, public square, etc.) as a major focus of design, and returns all urban buildings to an intimate scale (Jencks, 1981).

Enthusiasts insist that their scenario of small-scale neighborhood experiments is eminently cost-effective, given its expected savings in the energy, transit, and cost-of-living outlay of participants. Much is also made of the enormous, barely touched potential of "sweat equity". For example—

Baltimore's home-steading program, the nation's most ambitious, has preserved scores of blocks of dilapidated but essentially sound and potentially elegant 19th century red-brick row houses—something of a city trademark. For a $1 purchase price per house and the promise of sweat equity private citizens are restoring historic neighborhoods.... There is a similar program of shopsteading, whereby businessmen are encouraged to salvage old stores.... The ferment has stirred the spirit of the city's neighborhoods, 100 of which now mount their own festivals [Demarest, 1981].

In the midst of a post-affluent transition, then, the neighborhood scenario appears an especially attractive option to those who worry that worst times may lie ahead for many of the nation's cities, especially those that are energy and transit inefficient.

In the last twenty years of this century we will face an extraordinary need for experimentation, creativity, and invention. This is best done on a small scale. As James Madison once observed, a diversity of experimentation on the local level can only lead to progress. If the experiment fails, no great harm is done. If it succeeds, the experience can be rapidly transferred and adapted to other locales. It is this need for creativity and diversity that may be the greatest justification for neighborhood self-reliance [Courrier, 1980].

CONSERVATION CITY: ENERGY NEEDS

The Conservation City scenario is somewhere between the technological optimism of the Wired City and the joyful frugality of the Neighborhood City. In this scenario recognition is increasingly paid to the fact that human settlements have always been heavily influenced by their major source of energy:

> Today, as we move into the post-petroleum era, we are likely to see the birth of new kinds of communities that are influenced as much by new energy supplies and technologies as earlier cities were affected by railways and automobiles. This strikes fear in certain hearts, but hope in others [Corbett, 1981].

Certain anxious urbanites expect insufferable crowding and incompatible mixed use of the city block: An energy crisis may lead panicky people to crush together for warmth, even as business firms rush to relocate close to one another to reduce costly shipping charges. Other more hopeful urbanites, in contrast, envision a creative and humanistic response to any serious energy shortfall: People are thought likely to reuse urban space to minimize energy costs, even while maximizing style of life gains possible in a conservation framework.

Those who want urban sprawl curtailed indict it on many counts, the prime among which focus on the possibility of sweeping misdirection:

> There has been a growing concern that sprawl is upsetting the ecology, sapping the vitality of large and small cities alike, and devouring farmland at a dangerous rate. Today, there is new urgency, over and above these other still valid reasons, for containing sprawl. It has become one of America's notorious wasters of energy...by making mass transit too expensive and even unfeasible in many urbanized areas...by making commuters excessively reliant on the most energy-burning form of transportation, the private automobile. It leaves unused much of the capacity of schools, hospitals, fire stations, and other urban infrastructure that still has to be maintained. It requires the duplication of the same infrastructure outside the city. In the private sector, it hastens the decay of existing centers and fosters new ones far from where people live and work [Reuss, 1980].

In the place of sprawl would go unprecedented planning control, new restrictions on the leeway of private developers, urban densities as great as those in Montreal's Habitat (or in Hong Kong or Tokyo), and, the relegation of sprawl to the history books. In this scenario cities are urged to switch from horizontal to vertical growth, from sprawl to building into sky, from low-density to high-density, from the conventional to the mind-stretching, and from the familiar to the novel in all urban growth possibilities.

To be sure, far less exotic support measures exist, and they serve as useful ballast in the campaign of Conservation City enthusiasts. These down-to-earth prescriptions (no pun intended) urge the immediate cessation of sewer-permits and other indispensable government supports for sprawl development outside the city. Similarly, they urge that any new building be restricted to filling-in unused land (and vertical) areas in the city. Favorable attention is paid to zoning and tax incentives that will promote creation of higher-density shared quarters (nonfamily units, wherein unrelated people can live comfortably together).

With regard to the latter, King reports in the New York *Times*:

> Although there are echoes of the counterculture communes of the 1960s and 1970s, today's joint living arrangements are more of the checkbook than of the spirit. They also derive from changes in the social fabric: People marrying later, a rising divorce rate, the need for "family" support in the absence of a traditional family, and a society that accepts unmarried, unrelated men and women living together.... The elderly, often on fixed incomes eroded by inflation, many of them victims of high housing costs, are moving in together.... When seniors get together, they free up housing, so there is a double benefit [King, 1981].

Still other prescriptions seek to severely reduce the city-dweller's reliance on auto use, as this is expected to free up city land now reserved for street and parking use. At issue is whether or not a strong federal mass-transit policy can soon be won to help restore city transit systems to public respect.

Given images readily conjured up by critics of a Conservation City, images of an inhuman anthill or spirit-crushing termite mound, the controversy surrounding this futuristic, Buck Rogers-like scenario illuminates many open questions in urban affairs and urban policy making.

Critics, for example, scoff at the notion that a Conservation City means energy savings, as any conservation achieved through the use of multiunit buildings might be offset by the energy required to transport food into such a densely populated area, and, transport waste out of it. Proponents rebut by suggesting that a Conservation City scenario might draw on the local food-production focus of the Neighborhood City scenario—along with its far-sighted approach to a local recycling waste effort.

Other critics are quick to charge that a Conservation City will impoverish human life with its lack of private or semiprivate open space:

> Each of us needs a bit of the natural at home; a piece of earth where we can enjoy the sun, wind, rain, and growing things in privacy, and where we can fully interact with nature—not merely observe and experience, but dig, plant, prime, and harvest. This is especially true for those of us (the majority) who do not work on the land for our living. Public green spaces between multistory housing complexes are no substitute for the private or semiprivate yard [Corbett, 1980].

Conservation City defenders insist, in turn, that rooftop gardens, verdant balconies, and apartment gardens could flourish everywhere, thereby supplementing public green spaces, which would be more frequent and more imaginatively sited than true of anything at present.

Some critics are persuaded that advances in this scenario will only come at special cost to urban have-nots. In rebuttal, proponents insist that Conservation City is not one of no-growth, but rather, of directed growth. Have-not urbanites could profit from the cost-savings of energy conservation programs (retrofit of existing dwellings; upgraded requirements in new construction), even as they also gain from the guided, steady expansion of available housing stock and from obvious job creation possibilities. Other critics suggest that the higher densities may require new forms of social control and behavior modification. They urge creation of an "urban code of humanistic conduct" to help guide urban coexistence.

It is obvious that a modest conservation scenario is already unfolding in many cities, aided in part by the new tax advantages associated with historical renovation and presentation. Time, technological advances, and energy costs all seem to be working to this scenario's advantage, much to the excitement of its proponents.

INTERNATIONAL CITY: GLOBAL VILLAGE

The International City scenario is readily apparent in many American cities, especially those which historically have served as point-of-entry cities (New Orleans, New York, Boston, Miami, San Francisco).Other American cities, however, will have to take aggressive steps to insert themselves into the merging world system of cities (Dallas-Fort Worth, Atlanta, Knoxville, etc.).

Perlmutter (1979) has developed and refined an International City scenario which might be applied to any city. He contends that "in a world which makes 'instantaneously neighboring' technologically possible, the challenge is posed to the city, any city in the world—Do you want to belong to the world?—Do you want to recognize your linkage with the world?—Would you like to establish some roles that you would like to take in the world?"

Based upon his experience of years of international living, and his recent part in preparing a scenario for the future of Paris (at the invitation of the French Government), Perlmutter has outlined an international city scenario for Philadelphia.

First, to help guide this scenario, Perlmutter prepared a definition of the goal:

> An international city is a city where ideas, people, and products circulate with relative freedom, where people from other countries are given respect and a sense of dignity, are seen as individuals, and where there is a willingness to bring to that city the best of the world, including what the city itself has, building on the city's own strengths.

Given the city's many ethnic and nationality groups, and their links to nations abroad, Perlmutter is encouraged to think that Philadelphia has a strong foundation for grass roots support of his scenario. He also makes much of the human scale, the person-to-person hospitality for which the City of Brotherly (and Sisterly) Love is well-known—and which can help internationalism from being overwhelming.

To help get there from here Perlmutter recommends seven major steps: Philadelphia must immediately work hard to—

(1)   Develop information systems and export potential: This might take the form of a trading company to locate foreign markets for Philadelphia products and services. Special attention would be paid to ways for small and middle sized businesses to learn about their world markets.

(2)   Build up a regional financial center: This might involve helping local banks begin to act as a network, and also seek new foreign branch banking.

(3)   Attract foreign direct investments: This might involve wooing foreign business likely otherwise to locate elsewhere in the United States.

(4)   Create world-wide recognition: This might take the form of cultural events that build up the city's image and capabilities as a center for world events.

(5)   Reorient job training programs to the future: This would seek to match the vocational education process of minorities, unemployed Philadelphians, and young people to the nature of skills which are going to be needed—as these may be known abroad and to global companies.

(6)   Protect the "League of Nations" ambience of old neighborhoods: This might mean protecting the status quo—"If the communities are diverse ethnically, allow them to remain diverse; if they are homogeneous and that's the way they want it, then permit them to be homogeneous, still with an international character to the community."

(7)   Create an international city hub: The Delaware River area, at Penns' Landing site, is mentioned in this connection. It could include "a world trade center, with hotels, smaller buildings, and apartments, and the Penn's Landing Museum which could become an international exhibit center. This will be a place where the world comes to Philadelphia, a place where Philadelphians can meet the world" (Perlmutter, 1979).

He projects that if even half of these programs were successful, almost 30,000 new jobs would be created.

Such a set of scenario-promoting steps might be easy to envision in Philadelphia, which already receives almost a million foreign visitors a year. Other cities, however, need to come to terms with the steady emergence of a global economy and the pivotal role being played in its development by multinational business conglomerates. These cities can bid for and earn the branch location of dynamic foreign companies, as well as seek international linkages in education, health services, and tourism.

REGIONAL CITY: APPROPRIATE ALLIANCES

The Regional City scenario has been a fantasy for political scientists and political reformers since the beginning of this century. Their case for some form of regional governance has generally featured three elements:

(1)   Only the reconfiguration of existing political lines to meet new circumstances can provide adequate power to the parties involved.

(2)   Only the concentration of political power in a new regional framework can help Snowbelt cities match the superior influence in Congress of Sunbelt cities.

(3)   Only the rationalization of governance through the bold creation of regional authority will make optimum use of strategic new

methods (planning tools, computer modeling and simulation, forecasting aids, etc.)

In addition, some proponents speak to a still broader regional view:

No longer is it necessary to have a single, viable growing heart. Today's urban systems appear to be multinodal, multiconnected social systems in action...core-dominated concentration is on the wane; the multinode, multiconnection system is the rule.... The essence of the new urban system is its linkages and interactions, as shaped by changing modes of transportation and communication [Beny and Kasarda, 1977].

Given such a radical and revolutionary shift in urban ecology, conventional thinking about metropolitan growth and change may be in need of substantial rethinking, reevaluation, and reshaping. As they indicate, this "fundamental transformation of American urbanism has yet to secure a new definitive theory to match the conventional wisdom of the first half of the 20th century."

This alternative regional perspective is less concerned with changing the costly crazy quilt of quaint political jurisdictions than with developing a new mental model of a region's common fate. It has American urban regions moving from a concentric zone model to a multiple centers model. The former involves promotion of a strong commercial/industrial core in a city, with rings of bedroom suburbs surrounding it. The latter, in contrast, would have a smaller center city, one far more service oriented, possibly even a Conflict City variation. Industry and services would develop extensively in its surrounding area.

Radically different policies are required to promote either of the alternative Regional City models. Support for the concentric zone model entails improvement in public transportation, better employment opportunities (including blue-collar work) in the immediate outlying area, more jobs in Center City, less crime and better schooling (so as to encourage city residence), and increased investment in the city by public and private sources.

Support for the multiple centers model entails permission from suburban areas to allow growth of industry, investment in suburban industry and commerce, public policies to develop suburban jobs, and a leveling-off of gasoline and auto use costs.

With the development of a multiple centers model devoid of changes in political boundaries, it is a question of which institutions are able to create a regional forum or alliance able to aggregate talent, stimulate a common vision, and coordinate necessary decision making with respect to public and private investments.

Critics, of course, persist in dismissing all of this as so much wishful thinking. They argue that we have about as much regional cooperation at this time as we are ever going to, and that the opposition of local politicians to making any further concessions to regionalism cannot be exaggerated. Proponents retort that changing realities fairly demand new gains for new modes of governance, and cite the progress being made by on-going experiments in regionalism in Indianapolis, Jacksonville, Miami, Nashville, Toledo, and Minneapolis-St. Paul (where, in this last case, the Twin Cities have devised a tax-base sharing scheme that coopts the suburbanization of population and industry).

Gross (1967) urges us to move beyond being "still transfixed by the mirage of metropolitan government" and to invent instead new forms of twenty-first century regionalism. Such new forms of regionalism are likely to require significant commitments from progressive corporations, local foundations, universities, and hospitals, as well as from far-sighted public interest groups. An appropriate alliance would have to be formed.

LEISURE CITY:
THE REALITIES OF POST-INDUSTRIAL LIFE

The Leisure City scenario is potentially the most variegated of all the potential patterns of urban change. Retirement communities (Lake Havasu), recreation centers (DisneyWorld), and historical learning sites (Williamsburg) are all single-track manifestations of the elements embodied in this scenario.

A response to two different societal conditions—the prolongation of adolescence into the young adult years (22 to 28), and the expansion of longevity after retirement at 65—the scenario relies on a pleasure-centric reorientation of time expenditures.

Barol and Gerstl, two members of the Philadelphia task force that explored the seven urban scenarios, presented a four-element scenario which included these factors:

(1)   Work will become more impersonal with people spending their days typing data into machines while talking to each other over intercoms. They will commute either alone by car (a small one, at least) or aboard a train where people sit afraid of the person next to them. In short, people will spend their workday practicing exclusionary dynamics which should, according to the rule of

opposites, leave them ready for a leisure activity in which they can interact with other people.

(2)   Increasing technology will soon lower the cost of even the most sophisticated home video wonder. Gaming of this sort becomes more than an activity, it becomes a status mark—that is, until everyone has a home video center. After the saturation point is reached, then the time spent at this should level off.

(3)   A seeming contradiction in the wake of the burgeoning high technology leisure market is the present day back-to-nature, health-oriented attitude toward leisure. Jogging through the woods has often been prescribed as a cure for the tensions resulting from a day at work. Other forms of outdoor sports from hiking to softball aid this rebellion against the dominance of the modern world in people's lives. Neither sex nor age restricts people from these activities, as prescribed restrictions are falling left and right. Since there is little user cost in any of these activities, the public must ensure the continued availability of outdoor places fit for recreation.

(4)   Finally, as the workplace becomes less entrepreneurial and more managerial, emphasizing conformity over risk, people will seek activities in which they can perceive themselves as risk takers. Popular future activities, as well as those present now, will be those which make the participant the decision maker, risk taker, star player who becomes totally absorbed in the activity. (Look at the faces of those seated around the blackjack table at a casino. Activities that will duplicate this intensity will be successful in the years ahead.)

People, in short, are likely to seek more intense experiences; they will look for activities likely to make them feel different from their routine selves; they will seek personal growth; they will seek participation; and, they may want to socialize often and intensely.

The prospects for a Leisure City scenario are boosted by its job-generating, economy-vitalizing possibilities. A booming hotel business and the opening of many new restaurants, for example, provide semiskilled workers with fresh opportunities in the urban job market and with a cash flow that can partially alleviate the tax toll of the IRS. As this cash flow is generated by an influx of tourists, retired people, and students, it also affords a certain welcomed vigor to local banking institutions.

Any creative and dynamic use of leisure time and life-long learning opportunities in the society of the future will have both its urban and exurban manifestations:

In deciding how to use their free time, people have the opportunity to practice the skills of future-thinking: examining alternatives, com-

Table 4.3

Scenarios	Implementation Factors
I. Conflict City	Achieved by drift, inaction, and response to crisis; opposed by pro-action residents, but supported by civic cynicism, lethargy, lack of political vision, and class and race hostility.
II. Wired City	Promoted by commercial pressure on competitive firms and the lure of vast profits in new information services; opposed by technophobes and companies vulnerable to telecommunication breakthroughs.
III. Neighborhood City	Promoted by historic appreciation for small-scale community advantages; opposed by modernists who perceive the city as a single operational entity; opposed also by those who see neighborhoods as narrow bastions of self-defeating ethnic and racial isolation.
IV. Conservation City	Promoted by ecology and environmental conservation pressures and partisans; opposed by cultural hostility to "behive" living arrangements, and by traditional pride in the right to wide options in land development.
V. International City	Promoted by commercial pressure on competitive firms and the need to secure new jobs for local labor force; opposed by a provincial culture and a tradition of urban insularity and enthocentrism.
VI. Regional City	Promoted by recognition of inadequacy of city resources to meet problems with larger scope, and by the need to achieve economies of scale and central political direction; opposed by boosters of traditional political boundaries.
VII. Leisure City	Promoted by a steady contraction in the average work year, and by a steady rise in joblessness, and in dual income, smaller, better off households, and by pro-recreation culture; opposed by a Calvinist Work Ethic culture, by those who see leisure undermining the productivity of the local workforce and by those who fear "the devil will make work for idle hands," as in the case of an undereducated permanent caste of unemployables.

paring costs and benefits, rating the probability of undesired or unexpected consequences, and considering how best to balance risk against reward. And in a world that seems increasingly to be controlled by competing interest groups and power blocs, and where the individual seems to have little voice, free-time planning may soon be viewed as the last remaining bastion of personal power [Jennings, 1979].

IMPLEMENTATION CHALLENGES AND REALITIES

There are perhaps at least four time-honored response patterns to urban change:

—Least Effort Drift, or the passive acceptance of adjustment to the dynamics of urban development;

—Crisis Response Efforts, or the mobilization of dramatic efforts to respond to an unanticipated emergency;

—Recognition of External Opportunities, or the exploitation of new profit making opportunities embodied in the utilization of new technologies or techniques;

—The Appeal of the Novel, or the intrinsic, synergistic appeal of potential innovations and other unrealized opportunities.

These response patterns might be used by urbanologists to help determine which scenarios to nominate for citizen promotion or active opposition, for new dialogue, and for tentative experimentation. Table 4.3 outlines some of the implementation factors associated with each scenario.

All of the implementation factors in the seven urban scenarios remain open to active manipulation by key actors in each American city. As these scenarios challenge us to *choose* our urban future, they demand that both the functional elites and an aroused citizenry explore and assess their potentialities—positive and negative—with every iota of knowledge and wisdom at our combined command.

REFERENCES

BELL, D. (1980) "Introduction," in N. Simon Minc, The Computerization of Society: A Report to the President of France. Cambridge, MA: MIT Press.
BERG, P. and G. TUKEL (1980) Renewable Energy and Bioregions: A New Context for Public Policy. San Francisco, CA: Planet Drum Foundation.

BERRY. B.J.L. and J. D. KASARDA (1977) Contemporary Urban Ecology. New York: Macmillan.

CORBETT, M. N. (1981) A Better Place to Live: New Designs for Tomorrow's Communities. Emmaus, PA: Rodale Press.

CORNISH, E. (1981) "The coming of an information society." The Futurist (April): 14-21.

COURRIER, K. (1980) Life after '80. Andover, MA: Brick House.

DANTZIG, G. and T. L. SAATZ (1973) Compact City: A Plan for a Livable Urban Environment. San Francisco, CA: Freeman.

DEMAREST, M. (1981) "He digs downtown." Time (August 24): 42.

EVANS, C. The Micro Millenium. New York: Viking.

FRIEDMAN, L. S. and A. Wildavsky (1979) "Improving the quality of life: television repair." Technology in Society 1, 4: 329-338.

GORDON, T. J. (1981) "The year 2050: reflections of a futurist." Exxon Magazine (Summer): 31.

GROSS, B. "U.S. seen as Transitional society with varying rates of social change." The Futurist (October): 77.

HAYES, D. (1981) "Preface," in M. N. Corbett, A Better Place to Live: New Designs for Tomorrow's Communities. Emmaus, PA: Rodale Press.

JENCKS, C. (1981) Language of Post-Modern Architecture. New York: Rizzoli International.

JENNINGS, L. (1979) "Future fun: tomorrow's sports and games." The Futurist (December): 418-492.

JOHNSON, W. (1981) Muddling Toward Frugality: A Blueprint for Survival in the 1980s. Boulder, CO: Shambhala.

KING, W. (1981) "Costs and social changes promote shared living." New York Times (July 28): A-12.

LASKA, S. B. and D. SPAIN [eds.] (1980) Back to the City: Issues in Neighborhood Revitalization. New York: Pergamon.

LEONARD, G. (1975) The Ultimate Athlete: Re-visioning Sports, Physical Education and the Body. New York: Viking.

LIPSET, S. M. (1979) The Third Century: America as a Post-Industrial Society. Chicago: University of Chicago Press.

MARTIN, J. (1979) The Wired Society. Englewood Cliffs, NJ: Prentice-Hall.

MENZIES, H. (1981) Women and the Chip: Case Studies of the Effects of Information on Employment in Canada. Montreal: The Institute for Research on Public Policy.

MOLTOR, G. T. (1981) "The Information Society: The Path of Post-Industrial Growth." The Futurist (April): 23-30.

MORRIS, D. and K. HESS (1975) Neighborhood Power: The New Localism. Boston: Beacon.

O'NEILL, K. (1981) 2081: A Hopeful View of the Human Future. New York: Simon & Schuster.

PERGLER, P. (1980) The Automated Citizen: Social and Political Impact of Interactive Broadcasting. Montreal: The Institute for Research on Public Policy.

PERLMUTTER, H. V. (1979) "Philadelphia: the emerging international city." Philadelphia, PA: LaSalle College (International Business Lecture Series).

REUSS, H.H.S. (1980) "Opening Statement," in Joint Hearings, Compact Cities: A Neglected Way of Conserving Energy, U.S. House of Representatives, December 11 and 12, 1979. Washington, DC: Government Printing Office.

RUSSEL, R. A. (1978) The Electronic Briefcase: The Office of the Future. Montreal: Institute for Research on Public Policy.

RUTTER, L. (1981) "Strategies for the essential community: local government in the year 2000." The Futurist (June): 19-30.

SERAFINI, S. and M. ANDREW (1980) The Information Revolution and Its Implications for Canada. Quebec: Communications Economics Branch.

SHOSTAK, A. B. (1980) "High tech and the future of higher education." Phi Delta Kappa (February): 18-20.

SMITH, C. (1981) "Visions of tomorrow: life in the information age." New Age (September): 24.

SMITH, D. (1981) "Info City." New York (February 9): 24.

TOFFLER, A. (1980) The Third Wave. New York: Bantam.

VALASKAKIS, K. "The conserver society: emerging paradigm of the 1980s?" The Futurist (April): 5-11.

VILLADO, A. and K. DYCHTWALD [eds.] (1981) Millennium: Glimpses into the 21st Century. Boston: Houghton Mifflin.

WINFREY, C. (1981) "Leading consultant foresees the 'Wired Society'?" New York Times (July 5): 1.

Part II

Demographic Realities:
Is Demography Destiny?

☐ IT IS TRUE TO SAY that the future is not an empirical reality; until it happens, it is only a concept. However, the present has many commitments and extensions that reach out and preempt aspects of the future. These preemptions include budget commitments, contract provisions, intergenerational obligations, sunk investments, and the like. Even the distant future (50 to 70 years from now) will be linked to the present through the people alive today. The graduating college senior of the class of '82 will be finishing a forty-year career (multiple professions?) in the year 2022. Twenty-five years of retirement in some Sunbelt community will bring him or her close to the year 2050.

Some of the demographic realities were discussed in the introduction. Others are offered here. A fifty-year perspective on new urban populations in the American West is outlined by Butler and Chinitz. Bloomberg and Martínez-Sandoval focus on the development of a new type of American urbanization based upon the growth of the Hispanic populations in the border regions of the Southwest. Rose presents an historical analysis of the emergence of large black ghettos and speculates about their future. Destiny or not, the demographic realities will provide for both substantial continuity and substantial change in the next 25 years of urban development.

5

Urban Growth in the Sunbelt

KATHLEEN BUTLER
BEN CHINITZ

□ WESTERN MIGRATION HAS STIMULATED writing and thinking for as long as the first settlers in America realized what a vast land they were inhabiting. While the westward movement is not a new phenomenon, the issues surrounding westward growth are somewhat new. In particular, the issues surrounding the cities of the West and what they will mean for the people who move to the West and the country as a whole are of interest as the West continues to grow and attract a greater share of the country's population and industry. Certainly, as we watch the eastern and midwestern cities deal with the diminishing industries and the population out-migration that lead to near financial collapse, it makes us wonder whether the cities of the West now in their adolescent stages will become sparkling and vital cities of the twenty-first century or if they will reiterate the problems that plague today's cities. The Western frontier has been seen as a national safety valve because when things went wrong in other parts of the country the West was there as an alternative. While the wide-open spaces in the West may still be an attraction as a safety valve, the cities of the West are growing up and are not likely to be replaced by new towns growing into large urban areas. Concern over these cities, the urban centers that 100 years from now will stand as monuments to the twentieth century, is not unwarranted.

The eight western states—individually and as a group—are projected to grow much more rapidly than the rest of the United States in both population and employment over the next fifty years. Absolute numbers are shown in Tables 5.1 and 5.2. Growth measured against the 1978 base, is shown in Tables 5.3 and 5.4, and here the contrast to the remainder of the United States is evident.

Table 5.1
POPULATION PROJECTIONS: WESTERN STUDY REGION

	1978	1990	2000	2030
Arizona	2 353 827	3 030 226	3 442 227	4 441 599
Colorado	2 670 407	3 480 477	4 042 031	5 224 921
Idaho	877 539	1 064 497	1 193 915	1 416 097
Montana	784 553	853 301	925 564	1 130 563
Nevada	663 484	1 021 562	1 347 957	2 038 967
New Mexico	1 212 202	1 421 938	1 536 589	1 792 284
Utah	1 387 482	1 670 461	1 904 189	2 356 694
Wyoming	423 542	592 147	688 694	825 593
Region	10 373 036	13 134 609	15 061 166	19 226 718
U.S.	218 050 814	242 978 994	259 844 996	299 816 988

SOURCE: BEA/OBERS, 1980.

The net impact of the more rapid growth of the region on its share of U.S. population and employment is substantial. To state it one way, fifty years of disparate growth rates will move 2.7% of the nation into the region. In terms of the year 2030, that translates to 5 million people living in the region rather than elsewhere in the nation. Furthermore, the extra 5 million is a healthy share of the 19 million population projected for the region by 2030. Thus, despite the decline in absolute growth rates, the extra growth is impressive.

Both population and employment growth of the several sates will be similar except for Nevada, which will experience above-average growth, and Montana and New Mexico, which will experience lower growth than the region. But even these states grow faster than the nation.

CONTEXT OF WESTERN URBAN DEVELOPMENT

What is the context in which this western urban development will occur? Many of the factors that shape and have shaped cities in the past will continue to shape the new cities of the future. Location near commerce and trade, a historical precondition of urban growth, is one factor that will be important in the twenty-first century, and the issue is the types of industry and business that will be dominant. Natural resources development will be one of those industries and with the wealth of natural resources in the West, the cities that are able to

Table 5.2
EMPLOYMENT PROJECTIONS: WESTERN STUDY REGION

	1978	1990	2000	2030
Arizona	1 045 571	1 394 530	1 575 900	1 816 650
Colorado	1 351 434	1 893 309	2 192 762	2 606 525
Idaho	412 480	531 304	586 436	671 276
Montana	358 051	421 748	467 931	545 655
Nevada	393 361	614 324	770 698	1 004 607
New Mexico	518 037	556 387	725 762	815 594
Utah	582 742	815 096	954 417	1 154 120
Wyoming	225 725	326 609	365 054	412 633
Region	4 887 401	6 553 307	7 638 960	9 027 060
U.S.	100 458 447	121 183 000	130 076 000	139 990 000

SOURCE: BEA/OBERS, 1980.

Table 5.3
PERCENT RELATIVE POPULATION GROWTH: WESTERN STUDY REGION (WSR)[a]

	1978	1990	2000	2030	2030 (OBERS+)
Arizona	100	129	146	189	
Colorado	100	130	151	196	
Idaho	100	121	136	161	
Montana	100	109	118	144	
Nevada	100	154	203	307	
New Mexico	100	117	127	148	
Utah	100	120	137	170	
Wyoming	100	140	158	195	
Region	100	127	145	185	192
U.S.	100	111	119	137	

provide the financial and business centers around which that development will occur and prosper are going to be in the forefront of urban growth. It is these cities that will approach the scale of eastern cities as we know them today.

Table 5.4
RELATIVE EMPLOYMENT GROWTH: WESTERN STUDY REGION[a]

	1978	1990	2000	2030	2030 (OBERS +)
Arizona	100	133	151	174	
Colorado	100	140	162	193	
Idaho	100	126	142	163	
Montana	100	118	131	152	
Nevada	100	156	196	255	
New Mexico	100	107	140	157	
Utah	100	140	164	198	
Wyoming	100	145	162	183	
Region	100	136	156	185	191
U.S.	100	121	129	139	

[a] 1978 = baseline
SOURCE: BEA/OBERS, 1980; OBERS +, Mountain West Research, 1981.

Another likely pattern of development will be the extraction of natural resources in the rural areas of the West with some medium-size urban growth (50,000 to 100,000) in support of the site-specific trade requirements of the industry. Smaller communities (under 50,000) will also grow in support of more direct needs such as residential location. These communities, while not specifically company towns, are very likely to reflect the nature of the industry in the area, be it oil shale, oil and gas, or coal. Commerce in these towns will be directly related to the vitality of the industry nearby and the economy of these towns will fluctuate with the market for the product that dominates their area.

On the other hand, the larger cities of the West, those that approximate the scale of the eastern cities of 1 million in population or more, will be more diverse in their economic base since they serve not only the needs of the resource development in the hinterlands, but also develop their own industrial base. Before discussing specific sources of growth it is necessary to describe the context in which western cities will grow, highlighting the significant differences between the twenty-first century context and the nineteenth and twentieth-century context in which U.S. cities as we know them today grew and flourished.

The cost of energy in the twenty-first century will have a significant bearing on western cities. Unlike some eastern cities, whose form and function are well-established and are changing only on the margin, new western cities have the opportunity to adjust their form in light of the costs of energy. This is not to say that western cities will adapt to or respond to actual or potential high energy costs. However, as an initial planning step city administrators and public policy officials are paying lip service to the need to plan around the increasing energy costs. How these high energy costs will affect western cities is not entirely predictable; however, some areas that have been affected are (1) transportation patterns and (2) related residential and commercial development. In the same way that suburbs grew out of the industrial movement to the outlying areas of the older urban areas, they will continue to move to the outer areas of western cities. What is likely to be different is the acceptance of multiple activity centers in the western cities, and the corresponding planning for all types of development to occur in conjunction with those multiple centers. Energy costs will have and have had a positive affect on the construction of housing in these western cities. The available sunlight and solar technology, as well as other conservation measures, have become features offered to the average home buyer and are becoming expected housing features.

The explosive growth in communications and communications technology will also play a significant role in the development of western cities. One measure of a big city in the past was the concentration of businesses in the central business district (CBD). As the technology grows and is disseminated, the connection between the concentration of businesses in the CBD and the size and sophistication of a city will diminish. As financial institutions are decentralized, newspapers reduced to microwave transmissions, and industry dispersed throughout the metropolitan areas of western cities, the incentives and rewards for a central city location will be lost. In their place more mechanized and interpersonal forms of communications will become the standard and the value of interpersonal communication will become secondary.

Western cities will also grow during a time of fiscal retreat and a general distrust of government. Local infrastructure costs will have been on the rise and will continue to exasperate local officials as the lack of federal, state, and local financial capacity plagues these western cities. Even though the federal government was certainly not a major actor in the growth of the cities of the East and Midwest during the nineteenth and twentieth centuries, popular expectations were commensurate with the level of facilities and services provided. Public safety was seen as the most essential service to be provided and as long as the garbage was collected a citizen really had nothing to complain about. During the mid-twentieth century the public became accustomed to a broader range of facilities and services. Not only was it important to have police protection, but it was also necessary to have emergency services at the touch of a 911 number. Water availability was no longer as valued as was water quality, and of course, wastewater quality. Air quality, always considered a free commodity, has become a battle ground for those trying to define the problem and local government administrators trying to solve the problem. In the wake of restrictions on local government taxing powers and corresponding limitations at the state level, not to mention withdrawal of federal funds, western cities are faced with the need to provide a level of public infrastructure that not only meets twenty-first century expectations, but that also meets the demands of the projected population growth. Complaints of deteriorating subway systems in New York may be mild compared to complaints voiced in the typical western city of the twenty-first century where facilities and services were not provided when the funds were available from the growth taking place. As we can see in our eastern counterparts, periods of population decline and economic decline challenge the best financial conditions to maintain and repair, not to mention build new infrastructure.

Western cities will also grow in an environment that reflects the values and attitudes of a relatively affluent society. As the national economy shifts to more advanced industries such as the high technology business as well as the service sector, the people who will fill those jobs will be more well-educated and better-paid than their predecessors on the eastern seaboard. While Horatio Alger stories will always persist, the immigrants to the west will be coming to fill white-collar jobs and already moving up the corporate ladder. Thus, concern for quality-of-life considerations are likely to predominate the thinking of the twenty-first century.

COMPONENTS OF CHANGE

The principal source of growth for any city is economic, and understanding what types of business and industry are likely to precipitate growth is key to understanding the nature of that growth. For western cities economic growth will be of two principal types: natural resource and high technology. These two sectors will provide the goods and services that will be exported out of the West and consumed either within the United States or abroad. Looking more closely at the natural resource sector, it becomes clear that the resources are outside of the cities and located in substantially un-populated areas. Even given projections of population growth for such labor-intensive efforts like the development of oil shale, major cities will not spring up around energy development in rural areas. Rather, existing cities that have established transportation and communication sectors as well as financial industries will provide the base from which industry in the region will be developed. What this means for city development is that the people who work in these cities serving natural resource development will be white collar, professionals with college educations in accounting, geology, engineering, lay, and a variety of careers that connote both good education and good salaries.

As recently as the 1982 State of the State of Colorado address by Governor Richard Lamm, high technology has been identified as the promise of the future, and in fact western cities are ready, willing, and able, and are going after high technology firms. What is it that makes western cities favorable locations for this industry? High tech firms may be labor-intensive in their own right, and given the general nature of western cities, these firms are able to meet their labor needs for all occupational levels. For recruitment purposes these cities are able to offer a lower cost of living than either the east or west coasts, a favorable climate, relatively good educational institutions, and salary levels that have not yet been driven up by the voracious competition in the larger cities. Additionally, assembly-line jobs can be filled by a number of lower-skilled workers, not the least of which is the migrant workforce arriving daily from Mexico.

Another incentive for high-tech firms to move to the western cities, aside from the potential and existing labor force, is the general cost of doing business. Land values in the West, while rising rapidly, have yet to reach the levels seen in California's Bay Area and Manhattan. Thus, firms are interested in moving to these cities to establish their campus-style operations and possibly to reap the benefits of land value appreciation at some point in the future. Some firms have made conscious decisions that they no longer want the large concentration of employees in one area but are opting for several medium-sized facilities within a general metropolitan area. Western cities which have vast open space on their perimeters provide high-tech and other firms with the opportunity to diversify their locations and yet maintain some geographic proximity.

Another key factor in the decision to move to western cities is the progrowth environment that is promulgated by western cities, not to mention their state fathers. Western cities are well-versed in the use of industrial development bonds, tax-increment financing, and other mechanisms and incentives to lure industry. In addition, western cities are also characterized by low tax rates and minimal development restrictions. Some backlash is occurring with regard to the lack of regulations in cities like Houston and Phoenix; however, the free enterprise spirit underlies public policy in these communities and will continue to be the main thrust into the twenty-first century. In fact, the desire by most cities within the metropolitan areas to annex in order to survive, a belief held by many city administrators and politicians, will generate substantial competition for the industrial and residential newcomers, thereby decreasing the interest in greater regulation of development.

Obviously, there are exceptions to this scenario. Boulder, Colorado has imposed substantial growth restrictions that have driven up the land value and caused substantial speculation. Growth management is still a cause celebre.. However, communities are now concentrating on the management of growth rather than on no-growth attitudes and policies. Growth-management policies on a metropolitan scale are also much more difficult to develop and implement than those that only require the acceptance of one governing body. Western cities of the future will not be isolated but will be parts of metropolitan areas like the great cities of the twentieth century.

DEMOGRAPHIC PROFILE

The people who will live in these western cities will be important to their future character. The population will be made up of those indigenous groups who are and have been in the West during the twentieth century, inmigrants, and the children of both the inmigrants and the indigenous population. Beginning with the indigenous group, the largest growing group in the future will continue to be the Mexican Americans, who in the 1980 census have recorded remarkable gains in the Southwest. This growth, unlike the Anglo population, is characterized by a high fertility rate; this has significant implications for the future composition of the population. Mexican Americans are having children at a rapid rate, while the Anglo population is having fewer children. As long as the discrepancy exists between the Mexican standard of living and the U.S. standard of living, equilibrium at the border will not be achieved and pressure will continue on the U.S. side.

There will be two types of migrants into the western cities. The first is the person looking for white-collar occupations, the aesthetics of the West, and a lower cost of living than the east or west coasts. In fact, it may happen that this person would be willing to compromise in their salary requirements as a result of the combined attractions of the West. While business will see the West as a good place to move for a combination of reasons, including relatively low salary levels, these same salary levels will be considered by prospective movers in the context of their expectations of the benefits of a western lifestyle. It also is not likely that lower salary levels will continue into the twenty-first century in western cities to the degree they persist today. Competition for the best and brightest with eastern cities has and will continue to drive up the costs of labor, reducing the differences between payscales.

The second type of person that will be more willing to migrate to the West will be the person who needs a job at any cost. Historically, much regional migration has been caused by a depression in one region forcing people to move to the regions where the jobs are. While many regional economists debate the most recent census trends concerning why regions are gaining and losing population and which will continue to do so, it is clear that as long as the western cities are seen as lands of opportunity migration will be directed to those cities. The picture painted so far of the white-collar, middle-class population of western cities must be tempered with the expectation that the unemployed industrial worker of the East and Midwest will move to these cities in search of gainful employment. It must also be understood that western cities, for the most part, have never been characterized as having heterogeneous populations or being melting pots, but as being more nearly homogeneous white, middle-class populations. Some exceptions to this exist. For example, Albuquerque has substantial Hispanic and Indian populations, as do smaller towns near older mining areas that attracted immigrants in the nineteenth and early twentieth centuries and maintained their ethnic mixture. Even though these exceptions can provide examples of heterogeneous populations, ethnic and racial groups will be more dispersed in the bigger cities, perhaps because they will be more affluent and will desire to live in suburban locations, as opposed to the inner cities.

BUILT ENVIRONMENT

As businesses and population move into western cities, offices, houses, and transportation will be required. What this environment will look like, and even more importantly how it will work can be compared to the eastern cities as we know them today, as well as to absolute values or quality-of-life considerations. Automobiles, at the center of western values, will contribute to the future urban form. Even though mass transit systems are well-established in most western cities, neither priority or funding exist to develop these systems to a level or capacity that will handle the majority of the urban population of the twenty-first century. Inhibiting the potential for a greater role for the public mass transit are the existing residential development patterns. These patterns can be characterized as low-density sprawling subdivisions. Thus, the residential development patterns and the urban transportation systems become a self-fulfilling prophecy: The density does not warrant the investment in mass transit and the residential development follows the highway and street systems that are dominated by the car. Considering the level of investment required to construct, operate, and maintain any form of mass transit, the lack of federal funding, state funding, and local initiative in the twentieth century are likely to limit the role of transportation in the twenty-first century to that of the automobile.

Another significant characteristic of western urban areas will be the influence of the open spaces and outdoor recreation abundant in the West. Eastern cities have been defined by the concentration of cultural facilities in their core areas. Western cities will also have concentrations of cultural facilities, but the concentration will be diminished by the readily available outdoor recreation opportunities. Unlike immigrants to eastern cities from Europe, where capital cities provided strong images for cities, western urban dwellers will not have the imprinted desire for cultural experiences. Philanthropists may abound in the West as the wealth of natural resources are exploited; however, the values of the populace will not lean toward the enhancement of centers of culture. Historically, for example, in many of the turn-of-the-century boom towns, an opera hall served as the center of attention. In today's boomtowns an indoor recreation center and a commercial movie theater are seen as necessary cultural facilities.

TROUBLE IN PARADISE

What does all this mean to the quality of life in western cities in the twenty-first century? Will they be more or less livable than their eastern counterparts? Will social harmony be achieved? Will the environment and aesthetics of the West be maintained and enhanced? And finally, will the cities of the West show that we learned from the growth and decline of eastern cities? Western cities are the urban frontier of the twenty-first century, and how that frontier is settled will be the challenge at the close of the twentieth century.

The climate and geography of the western states is such that the region is generally short of water, sparsely populated, held together by limited transportaion facilities, and has an economic base of diversified agricultural, recreational, manufacturing, service, and extraactive industries. Much of the region is provided with enormous recoverable energy and mineral reserves which will provide the impetus for economic and demographic growth. Energy production, particularly coal and oil shale, is projected to expand considerably in the next fifty years as the nation looks to the region for its domestic energy supplies.

Although the West contains vast mineral and land resources, it is nevertheless a region shaped by resource competition and scarcity. In 2030 competition for scarce resources—water and land, in particular—will become a predominant factor in determining the economics and characteristics of development. The availability of water, and in some cases land, will be limiting factors in the development of the economic viability of the region.

CRITICAL ENVIRONMENTAL ISSUES

The most important environmental issues will surround the disposal of enormous volumes of surface-retorted spent shale, increased urban and industrial nitrogen oxide (NO_x) emissions, and the disruption of aquifers by underground mining. Although their impact may not necessarily be the most severe in any particular area or across the western region in general, they will present the greatest challenge to mitigation efforts, for technologies and innovative policy options are not presently available to avert them.

Other critical environmental issues will be associated with the predicted western development. The degradation of water quality could become a serious problem as aquifers are contaminated by waste from both energy and nonenergy industrial facilities and from uncontrolled runoff from increased urban activity. Even assuming that sulfer oxide (SO_x) emissions will be controlled at levels required today for new sources, SO_x emissions will increase over the next fifty years. Increased SO_x emissions will have a noticeable impact on visibility, which is with few exceptions relatively good throughout the region, and will reduce the intrinsic scenic value of many of the western parks, monuments, and wilderness areas. Together with increased nitrogen NO_x emissions, these acid-forming pollutants could lead to acid rain in portions of the Rockies, which could damage fish populations in the high lakes. The increase in NO_x emissions over the next fifty years will produce a substantial increase in regional brownish haze, especially during winter months.

Technologies to control NO_x emissions are in the development stage now. It would not be unreasonable to expect that effective technologies to control NO_x emissions from coal-fired power plants and synfuel facilities will be available well before 2030. SO_x emissions from coal-fired power plants could be reduced beyond current regulatory requirements by employing existing control technologies. A total systems approach may indicate that it will be more economical to control NO_x and SO_x emissions by employing advanced energy technologies. The technical and economic feasibility of these advanced technologies should be known by the turn of the century.

In urban areas it is expected that the levels of carbon monoxide and possibly of ozone will be lowered as cleaner-burning automobiles constitute an even-larger proportion of the fleet. Particulates and NO_x emissions, on the other hand, are expected to increase unless automobile controls for these pollutants are improved. In the absence of any improved controls, urban haze will probably expand and slightly intensify.

In some areas surface water supplies will not be sufficient to meet the demands from both urban and industrial developments. New sources of water supply will have to be found in conjunction with the development of methods to reduce demand. For example, the use of wet/dry cooling towers at coal-fired power plants would consume only 25% of the water otherwise required by wet cooling towers. All new demands for water are projected to be for nonagricultural uses. It is likely that much of the new water demand will be met through transfer of agricultural water, which will thus bring about changes in the patterns of agricultural production and the introduction of new irrigation techniques. The diversion of water from agricultural uses will reduce the salt burden added to rivers and, consequently, will contribute to the improvement of water quality in those areas.

SOCIAL ISSUES

Economic growth and expansion in the Western Region pose significant social issues for the indigenous population and in-migrants coming to the region for economic opportunities. One issue centers on whether the racial and ethnic minority groups will share in the benefits of the projected economic growth. Some projections of growth in high technology indicate that highly educated and trained labor will fill the job needs. Potential disparities and tensions between those who benefit from the healthy economy and those who do not may develop over the new wealth. Whether any trickle-down effect will occur is open to speculation.

A second issue arises from the prosperity projected for the region,, that is, if high labor, land, and housing costs will eventually create the same disincentives for industry as they have in other areas in the past. With the current slowdown in national mobility (which partly reflects the high costs of moving), the desirability of a western lifestyle may not be sufficient to overcome the high costs of relocation. On the other hand, the higher cost of living in the region will adversely affect those living on fixed incomes and disillusion those who have moved to the region who expect to enjoy a higher standard of living.

Finally, the loss of power by those residents who have lived a good part of their lives in the region may contribute to friction when the new, projected forces take the stage. The region's immigrants may bring substantial outside influence, attitudes, and values to the region, and these may not be compatible with existing regional attitudes and values. Moreover, to the extent that capital for economic growth comes from outside the region, the density of the West may no longer belong in traditional western sources of power.

We might also note the relatively slow growth of government employment as compared to total employment. This projection poses an interesting dilemma. Can the region cope with the impact of growth with a government sector that grows more slowly than the economy as a whole? As economic and population growth occur, the demand for public facilities and services typically grow. This potential growth of government may also introduce new forces and values.

SUMMARY

In summary, the overall projections for the region as a whole show a healthy margin of growth for both population and employment over national rates, but the absolute rates of growth will decline sharply in the future. Yet, in 2030 we will have a region with approximately twice as many people and jobs, at which time each job will be associated with three times as much output in constant dollars.

Underlying and shaping these economic trends will be a variegated collection of political alliances that will coalesce around specific issues, disband as common causes splinter, and regroup as changing issues, needs, and perceptions demand. The driving decision variables will be scarcity, sovereignty, and social stability, including the economic and political responses to a tightened supply-demand situation affecting basic resources; the quest of each participating level of government to assert and preserve its proper sphere of authority; and the tendency of the populace to resist and react to change and to ensure an enduring element of continuity in the status quo. What has occurred so far is miniscule when compared with what is projected for the future, particularly in the next fifty years.

Thus, as we look to the future we see a region in which the new post-World War II growth forces merge with the revival of intense interest in the region's energy resources to maintain the region's growth record as compared with the U.S. average.

While the spotlight may be on energy development in rural areas, we do not expect the character of growth in the region as a whole to be dominated by energy. Instead:

—The manufacturing and service sectors will outpace the growth of the energy sector, and the latter will increase only slightly as a share of total employment.

—States with no significant energy resources—Arizona and Nevada—will grow as fast or faster than the regional average. Urban areas throughout the region will continue to grow rapidly despite their distance from the energy boom, albeit their growth will be encouraged in part by the linkages between rural and urban growth.

—Most of the tangible impacts of energy development will be felt in both the immediate vicinity of the activity and in nearby communities. Some rural areas will, as sites for energy projects, experience very rapid growth.

—In these rural areas the greatest limiting factor under this scenario will be dominance of resource development itself. Resource development will take manpower from agriculture and other potential industry and thus, will inhibit broadening the economic base as the economy reacts to the development opportunities.

—A substantial share of the growth will be urban in character, pushing small towns and cities upward in the urban hierarchy and creating a tier of medium-sized cities (150,000 to 250,000) which may not have otherwise existed at all.

—If communities are determined to reach these new levels of development, manpower, materials, and money will be brought into the region at required levels. Resource (air and water) appropriations would also take place as needed within existing institutional constraints.

—The effects of energy development in urban areas will not be as significant to their economic development as in the region's more rural areas, since growth in manufacturing is the principal growth sector projected for the next fifty years. Urban areas will reflect a broader, more diversified economic base than will their regional rural counterparts.

The urban areas will develop complex new linkages. They will expand their entrepot role in the movement of people and freight. The increased demands of the hinterlands will expand their service center role. The decision-making jobs in both the energy and construction industries will be located in and near cities, and many of the new on-site workers will elect to live in nearby urban areas. Western cities in the year 2030 are likely to be remarkably urbane in their character.

6

This Hispanic-American Urban Order: A Border Perspective

WARNER BLOOMBERG
RODRIGO MARTÍNEZ-SANDOVAL

☐ FOR MUCH OF URBAN HISTORY major areas of many cities have been identified with particular ethnic populations and communities. It has been more unusual for a whole city to bear such an identity. In such cases it frequently has been the urban center for a region in which that ethnic population dominates. In Switzerland, for example, Zurich is German and Geneva is French, as Quebec in Canada also is French. Such cities and regions are not in fact ethnically homogeneous. What gives them their identity is the way in which the demographic dominance of one ethnic group has been extended into linguistic, cultural, and political domination as well. The United States has never had any major cities with such a character; but it may well experience a regional network of them in the Southwest as the Hispanic population and its cultural and political influence continues to increase into the early decades of the twenty-first century.[1] This analysis explores the bases for this possibility, notes factors that militate against it, and assesses probabilities in the areas represented by Map 6.1.

ETHNIC AGGREGATIONS AND URBAN STRUCTURE IN THE UNITED STATES

Migrations into and within the United States have militated against the prolonged demographic domination of any region by a single

AUTHORS' NOTE: *We wish to express our appreciation to our colleague in the School of Social Work at San Diego State University, Dr. Juan Ramon Valle, for very helpful and informative criticism and encouragement in the development of this analysis.*

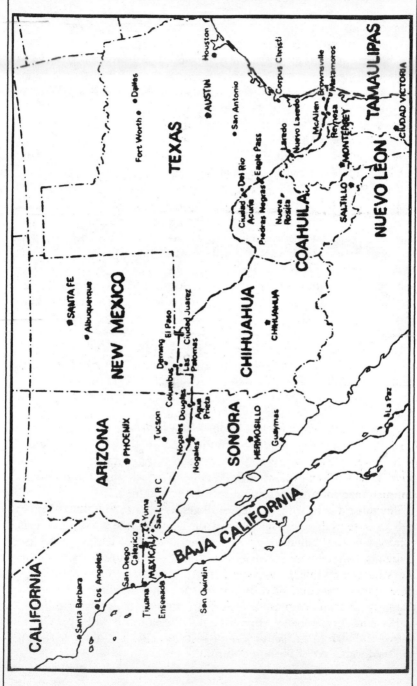

MAP 6.1 The United States-Mexican Region

113

ethnic group other than the majority. However, in some regions and cities the original WASP majority was demographically overwhelmed and politically diminished by the aggregation of European and "non-white" ethnic groups. The largest and most powerful of these groups have been able to impress their interests, cultural as well as economic and political, upon the overall patterns of the metropolis, and to use such bases for greater leverage in state and national affairs. The Irish and the Italians of Boston, the Jews of New York, the Poles and the Irish of Chicago, and the Germans of Milwaukee are examples from the first half of this century. Nonetheless, there were no Quebecs.

Ethnic subcultures were most strongly manifested through familial and neighborhood lifestyles and through institutions subject to substantial local community influence: churches, taverns, local stores, and the like, and the ward organizations of political parties. The suburban exodus of WASPs from the central city originated in part as a way of putting spatial as well as social distance between themselves and the growing influence of the white ethnics. But in recent decades the latter have joined the suburban movement in ever greater numbers, leaving behind aging residences and increasing numbers of non-white residents.

In spite of the much touted gentrification of certain very limited areas, many of the central cities are rapidly becoming dominated demographically, although not yet politically (and perhaps never economically), by Blacks, Hispanics, and the new immigrants from Asia and the Middle East. In a few instances central cities are becoming identified with one group. Newark, New Jersey, is now a Black city; Miami is rapidly becoming Cuban. By the second decade of the twenty-first century, there will be a large number of central cities in which Whites, whether WASP or ethnic, are a numerically diminishing minority, in some cases less than a third.

The pattern emerging in the Southwest is unique in both its history and its potential. Especially along the Rio Grande there have been some towns and small cities with a majority of Mexican-American residents, where Spanish has been as much the language of the shops and streets as English. However, the political and economic life of these places was controlled by the Anglos until recently. Now that is changing, even as the population of the border area grows rapidly. In California, Los Angeles (the city, not the county) will be Mexican before the turn of the century. Not just that city, but a whole region, will be dominated by Hispanic communities.

OLD CITIES, NEW IMMIGRANTS, AND ALTERNATIVE FUTURES

Barring a nuclear holocaust a high probability already obtains for certain aspects of the next half century. Five are posited as givens for the analysis that follows, and as a framework within which to examine probable and possible interactions of less predictable factors.

Economic

Within a world economy ever more dominated by multinational corporations and compacts, the United States will continue its postindustrial evolution (Bell, 1973; Henderson, 1978; Kahn and Phelps, 1979). The primary sector, recapitalized to exploit technological breakthroughs, will provide proportionately even less employment for the less-skilled. A steady application of computers and robots will continue to reduce labor force in the secondary sector, whose expansion will be stimulated mainly by population growth and will be constrained by resource conservation and some redefinition of the good life toward services and experiences rather than material possessions. The tertiary sector will thus continue to expand, especially through public sector and mixed private/public enterprises and delivery systems (Edmunds, 1980). The fastest growth will be in the now rapidly emerging quaternary sector (research and development, communication and knowledge transacting systems) with respect to investment and markets, but not labor force (Gappert, 1979; Molitor, 1981).

Regional and Urban Ecology

The completion of the expansive phase of development of the western regions, the industrialization of the South, and the revitalization of the older regions of the Northeast and Midwest will occur during these five decades. Regional differences will diminish substantially in terms of such indicators as per capita income, education, and the sector composition of the regional economies, although this does not preclude some waxing or waning of a region over a decade or two as technological, investment, population, and market shifts result in some period of advantage or disadvantage relative to other regions (Berry and Kasarda, 1977). Most of the Standard Metropolitan Statistical Areas (SMSAs) within the large urban regions will become "dual metropolises" (Bloomberg, 1979), with the areas outside the old central cities no longer dependent suburbias but increasingly independent urban conglomerations with all needed services, amenities, and employment opportunities for a large proportion of their residents (Colenutt, 1972). A limited, highly selective gentrification of the old inner city will establish residential islands close to those financial and mercantile areas providing occupation, shopping, and services for the well-to-do, most of whom will be Anglo and many of whom will still be commuters from the outer metropolis (Allman, 1978). The rest of the central city (some to a lesser degree than others) will become increasingly non-Anglo/white, with the worst of present poverty alleviated by more cost-effective welfare, and with a larger proportion of petit bourgeoisie and lower-middle-income households (jobs for the latter located in secondary as well as tertiary sector occupations) than in the outer metropolis (Downs, 1976).

Mexican Border Region Development

Whatever periods of turbulence may occur, and some could be quite explosive, Mexico will continue to industrialize. During the 1980s and 1990s, and perhaps beyond, migration into the northern tier states, investment programs by the federal government and private entrepreneurs, and growing interaction with adjacent U.S. markets and urban systems assure the relatively rapid development of the Mexican side of the border region and its increasing influence in national affairs, even though this growth also may be thoroughly exploited by U.S. business and industrial interests (Cárdenas, 1977; Fernández, 1977; Samora, 1975). An urban region encompassing Tijuana, Mexicali, and the lesser cities of Ensenada and San Felipe will be the first to reach developmental maturity and a high degree of interaction with the Los Angeles-San Diego-El Centro-Yuma region on the U.S. side. Others will develop later and to a lesser degree by 2030, but their growth, industrialization, and increasing interaction with U.S. counterparts is inevitable: Ciudad Juárez and El Paso where Texas, New Mexico, and Chihuahua intersect; the string of towns and small cities along the eastern stretch of the Rio Grande, anchored by Matamoros and Brownsville and probably aided more by the expanding economy of Monterrey than by the wealthy urban region to the north (Houstin-Austin-Fort Worth/Dallas) from which they have been systematically excluded; and eventually Los Ambos Nogales south of Tuscon, which provides a connection with Arizona's sunbelt expansionism.

Migration from Mexico and U.S. Demographics

"Discovery" of illegal immigration has initiated revisions of forecasts for both total size and composition of the U.S. population

(Huss and Wirken, 1977). Migration will continue at a high level from Latin America, especially from Mexico, which has contributed close to 70% of the total U.S. Hispanic population (Estrada et al., 1977). In spite of agitation to stem the flow across the border, powerful push-and-pull factors will continue at a high level for at least several decades. The drop in the Mexican birth rate since the family planning campaign was launched in the late 1970s has been mainly in the urban areas. Rural communities continue to send migrants to the cities and thence to the northern region and into the United States—a consequence of Mexico's surplus of unskilled workers unaffected by its increasing need for skilled labor (Martínez-Garcia, 1981a). Since those who will be joining the labor force in 1995-2000 were born prior to the family planning campaign (Coale, 1978; Martínez-Sandoval, 1981), the Mexican government's present population control programs, including its population relocation and retention plans for the turn of the century (Consejo Nacional de Población [CONAPO], 1980), will not lessen the population pressure on the Mexican economy. This situation allows for a continued migration that serves important capitalistic interests in both countries (Bustamante, 1978; Cornelius, 1978; Portes, 1977).

On the other hand, without the addition of migrants the United States could experience a labor shortage even before 2000 (Cornelius, 1981), especially for the many low-skilled, low-status jobs which will continue in spite of the expansion of new technology, attracting workers mainly from Mexico and Central America (Martínez-Garcia,1981b). Many will remain in spite of greater encouragement of a revolving door pattern; and those who become residents will have notably higher fertility than Anglos, both because of the young age of the inmigrants and because of cultural support for having additional children (Bean and Bradshaw, 1977), even though there may be a long-term, intergenerational trend toward convergence of fertility rates.

Sociopolitical Patterns

A three-tiered political system already evident in the United States will continue to constitute the basic structure of power, but a variety of elaboration and modest modifications are both possible and not very predictable. The greatest power over national policies will continue to be wielded by the elites (Domhoff, 1967, 1971), including the highest levels of the military and the federal establishment. But the peripheral extensions of their control and influence will have been contracted in many areas of secondary interest in the face of recurring efforts from below to assert claims in areas of special interest.

The second tier of power begins among the wealthy and well-organized below the elite level and extends into the upper levels of the broad middle class. A growing proportion will reside in the outer metropolises of the major urban regions and in the islands of gentrification within the central cities, but their interests and influence will tend to be regional and national expressions of occupational commitments and concerns for protecting their way of life, including the spatial areas in which it is pursued.

The third tier will be both the most heterogeneous and the least powerful except for local affairs: those whose incomes consign them to the older core cities; minority group members who have not been able or else have not chosen to move into the outer metropolis; most of those dependent upon transfers from the public sector of goods, services, and/or disposable income; poorer farmers on the spatial and marketing peripheries of corporate agribusiness; the less-skilled of the still declining proportion of blue-collar workers and their counterparts in white-collar occupations. Even the well-organized among those in this tier will be able to exert important influence at state, regional, and national levels only in coalition with segments of the second tier with whatever trade-offs such coalitions require.

Though these five factors, we believe, constitute a framework within which the future of urban areas in the Southwest will unfold, much remains uncertain: social movements, ethnic group coalitions, personal and public policy patterns affecting residential segregation, public and private investment in relation to labor force allocation by economic sector, and so on. The next step in this anticipatory history is to examine some of those less predictable variables most relevant to development of Hispanic central cities.

URBANIZATION IN THE SOUTHWEST:
THE HISPANIC DIMENSION

In so brief an assessment of the factors which will give hispanics a place in U.S. development distinct from that other immigrants have played, little attention can be paid to the rich diversity Mexican Americans display from one area to another in the Southwest, reflecting origins in Mexico, differences in social class, and unique histories of particular communities (De la Garza, 1981a; Knowlton, 1975; Nostrand, 1975). This reminder of that diversity must serve for all that follows in exploring six key factors.

SETTLEMENT PATTERNS

The consequences of immigrant concentration in certain areas are illustrated by Puerto Ricans in New York and Cubans in Dade Country, Florida. Those from Mexico, so often denied documentation, have been moving along the routes of low-wage exploitation followed by their ethnic predecessors, both migrant workers and those entering the secondary and marginal occupations of the cities (Cárdenas, 1977), joining established enclaves of linguistic and cultural familiarity. In spite of the resultant burgeoning of Mexican-American communities in the plains states and the Midwest, a disproportionate share have remained in the Southwest. Thus, in California between 1970 and 1980 total population increased by 18.5%: Whites by 5%, blacks by 33%, and Hispanics by 60%—a rate of increase surmounted only by the 140% increase of Asians. And by 1980 Los Angeles County had almost half of all the Hispanics in the state. Within the city they will be the largest single ethnic group by 1985—probably a third of its total population, or more. By 2000 Los Angeles could be like El Paso in 1980, with 60% of the residents Hispanic—and what will El Paso's percentage be in 1990? Or Brownsville's (77% in 1980)? Or McAllen's (81% in 1980)? It should also be noted that these heavily Mexican-American Texas border cities increased in total population by from a third to one-half between 1970 and 1980 (U.S. Bureau of the Census, 1980), a growth rate which is expected to continue at least through the eighties.

But will there be a second-generation reversal of the trend, as occurred with so many White ethnics, followed by continuing dispersion into the third generation, spreading Mexican Americans much more evenly throughout the United States and into the outer metropolises? The remaining five factors and U.S. racism both militate against such a reversal.

THE "OVERLAND BRIDGE"

Unlike any other immigrants who have come in large numbers, those from Mexico can get into cars and drive back. Moreover, most of them are well aware that most of the U.S. Southwest was Mexican territory less than 150 years ago and that it was taken from Mexico by military power. There simply is nothing comparable in the experience of the immigrants from Europe or Asia, or even from other parts of Latin America (Alvarez, 1971). The immediacy of the connection serves as a taproot for continuities of identity and culture. The "Mexican" of Mexican-American is also continuously invigorated by a substantial portion of immigrants who exercise any of the revolving door versions of residency in the United States, from annual round-trip migrations associated with employment to much more prolonged episodes in each country. For those living a day's drive to the border, holiday visits with friends and kin are easy to accomplish, and for those living in the border region an ordinary weekend, or even one day, may suffice. This does not mean that Mexican Americans necessarily have more of a dual loyalty than other immigrants have displayed. Indeed, these real connections provide more of basis for a critical view of the nation left behind than when an ethnic ideology is created around an idealized fantasy of one's now distant former or ancestral homeland.

THE BORDER REGION

The change in jurisdictions between Canada and the United States, while significant, contrasts strikingly with that between the United States and Mexico. We do not think of a United States-Canada border region. In terms of the everyday lives of those living along the boundary, and of the public and private corporate entities through which much of that everyday life is organized, the transition from one nation to the other is made so easily that it can almost be taken for granted. Thus, Detroit and Windsor share in the cycle of annual festivities, and the two cities are integrated into a single metropolitan area and urban region almost as fully as if they were both within the same nation. But to go from San Diego into Tijuana, in spite of the many "yankee" influences on the latter (so often criticized by Mexico City), is to cross a great divide involving culture, language, legal code, currency, governmental structure, political tradition, law enforcement—nothing matches well or dovetails right up to general level of economic development. Yet the flow of trade and people across this border, both legal and otherwise, has been creating many elements of a common urban area and will continue to do so (Appleyard and Lynch, 1974). With this there is a growing number of "border people" whose daily or weekly round of life straddles the boundary. As this cadre of residents grows in the decades ahead, they will help to create a thickening network of informal ties and an attitudinal atmosphere conducive to the development of transborder organizations and agencies in both the public and private sectors wherever there are twin cities. Harbingers of this future are already visible where a large Hispanic population on the U.S. side facilitates collaboration. Business groups in Matamoros and Brownsville engage in a mutual promotion of the tourist industry; a sewage plant in Nogales on the U.S. side serves the largely Mexican Nogales as well.

Increasingly throughout the next half-century the people and the communities of the border region will mediate between the large societies on both sides, their influence growing with the border region's general increase in population and economic vitality in spite of criticism in both societies that the "twin-ness" has been carried too far. But it is difficult to see what might lead to a real reversal of present trends, short of actual conflict between Mexico and the United States.

LANGUAGE, EDUCATION, AND CULTURAL IDENTITY

Language is the heartland of cultural identity. The common use of Spanish, albeit with differing dialects, links Puerto Ricans, Cubans, Mexican Americans, and other Hispanics in the United States. Assimilationists notwithstanding, the restoration and maintenance of that linguistic tradition commands increasing support both within and between the different groups of Hispanics as an affirmation, not of nationality but of cultural identity and group cohesion. To the extent that they succeed in having binguality implemented as social policy, they will establish yet another difference between their present and forthcoming experience and that of immigrants from Europe during the late nineteenth and early twentieth centuries. Until very recently public schools in the United States were used as instruments of assimilationist policy targeted upon "un-American" patterns of behavior, especially language (Leibowitz, 1971). Parochial schools served to insulate some members of the second generation against the WASP assault, preserving ethnic and language, as well as Catholic, commitments (Fishman, 1966). Bilingualism also was promoted as a core to cultural identity in German Lutheran parochial schools and by a variety of efforts to institutionalize and sustain Yiddish, primarily in the Jewish segment of New York City. Nontheless, since the gateways of for mass migration from Europe were closed, bilingualism has largely disappeared among white ethnics, along with much else that was subject both to deliberate assimilationist replacement and to the continuing erosion that so often accompanies individual and group social mobility. For the third and subsequent generations being "ethnic" has become increasingly a matter of intent with ideological and political connotations (Glazer, 1954; Glazer and Moynihan, 1963).

If this became the dominant pattern among most Hispanics in the United States, as it already is among some, then Spanish would decline as a public language coordinate with English in marketplaces, on public documents, and most of all in the school curriculum, where bilingualism would be constrained narrowly and minimally to aiding newcomers in switching over to English. But, as the anthology edited by Cotero and Hufford (1980) indicates, there is a growing cadre of intellectuals and practitioners dedicated to sustaining and even increasing bilingual education as public school policy in cities where children from Hispanic families are a substantial segment of the public school population, and further to establishing the legitimacy of Spanish as an official public language.

POLITICAL PARTICIPATION

The future of Spanish in the areas where Hispanics are concentrated is thus to an important degree a question of political power. Much of what happens in the politics of the Southwest during the next fifty years, with consequences for the nation as a whole, will hinge upon the rate of increase in political participation by Mexican Americans, and upon the extent to which their political activism is absorbed into and expressed through established parties and organizations such as the PACs of trade unions, professional associations, and business groups. The organization of farm workers by Cesar Chavez, high school student strikes (mainly in California) in the late sixties, the more recent Chicano movement on college campuses, the election victory in Crystal City, Texas, and subsequently in other border towns, efforts to organize La Raza Unida as a Mexican-American alternative to absorption into the Democratic or Republican parties— all indicate a growing politicization of Hispanics in the Southwest.

Nonetheless, political participation is likely to increase only slowly through the 1980s. A large proportion of Mexican Americans remain poor and poorly educated, inhibiting political participation. Immigrants from Mexico often bring with them an alienation from the political arena; and, of course, neither undocutented nor legal aliens can vote. But by 2000 much will have changed. One way or another the politically difficult problem of the undocumented inmigrants will have been resolved, and most of those now present or entering in the next five years who choose permanent residence will have been able to obtain citizenship by the beginning of the next century. Moreover, their children will be citizens; those born by the publication of this volume will be old enough to vote. Overcoming political apathy is already the most pervasive issue on the agendas of Chicano elected and appointed officials and community leaders (De la Garza, 1981b; Rubio, 1980; Torres, 1979).

By or before 2010 the already growing number of Mexican Americans elected to local offices will have become very large throughout the Hispanic Southwest, as will the proportion of Hispanics in government bureaucracies. The Spanish-speaking electronic network, both TV and radio, already multinational in origin and distribution of broadcasts, will have provided connections of image, news, and commentary informing and reinforcing Hispanic publics in the United States in a way never experienced by other ethnic or racial groups, contributing to the manifestation of the potential for political activism shown in the past and to overcoming the alienation from involvement produced by more contemporary experience. Finally, relationships between Chicano organizations and the Mexican government, presently in a stage of early development and uncertain direction (De la Garza, 1981a), will become more extensive and consequential in the decades ahead, adding greatly to the leverage Mexican-American groups can exert in dealing with Anglo decision makers in the second and first tiers of the political system.

ECONOMIC PARTICIPATION

A large proportion of Hispanics, like Blacks and other minorities, have been, are, and will continue to be confined to the lower, limited-growth segment of an increasingly dual economy in the United States (Averitt, 1968; Beck et al. 1978; Rose, 1981), lacking the income, fringe benefits, and security associated with the high-wage, unionized, and highly skilled strata of occupations. About 7.5% of Hispanics are in the professional and technical category, and about 5% are nonfarm managers and administrators. Almost a third of employed females are in clerical and kindred jobs, a quarter are operatives, and another fifth are service workers (U.S. Bureau of the Census, 1979). About 13% of the males are also service workers, a tenth are nonfarm laborers, a quarter are operatives (including transport), and a fifth are in craft and kindred occupations. To the extent that advancement in the postindustrial segment of the U.S. economy has educational requisites, this generation of Hispanics is clearly disadvantaged: only 16% of those 25 or older have any college education (about 7% have four years); 43% have no high school. Moreover, those of Mexican origin rank lowest for education—below Puerto Ricans, and far below Cubans and others from Latin America. However, Mexican-American incomes are higher than Puerto Ricans' although less than that of Cubans and others: About two-fifths of the Mexican American families had incomes of $15,000 or above in 1978, while somewhat more than a third had under $10,000 (U.S. Bureau of the Census, 1979). Large families—six persons or more—occur almost twice as frequently among Mexican Americans (19%) as among those from

Puerto Rico and Cuba (U.S. Bureau of the Census, 1979). But Mexican families, especially at lower-income levels, are characterized by the participation of everyone possible in the job market in whatever way possible, however menial and part-time, with great resistance to acceptance of public assistance.

Mexican Americans are displaying their strong impulse to parlay "kinship capitalism" into a proliferating variety of mercantile, service, and small-production petit bourgeois enterprises. It remains to be seen if in the cities they dominate demographically and politically they will be able to create a local structure of markets and occupations, facilitated by regulations and inputs negotiated with the second-tier decision makers at state and national levels, that is viable, albeit with less affluence for most residents than one will find in the outer metropolis.

Some economists doubt even the theoretical possibility of a local urban economy that, although necessarily articulated with the regional and national economic systems, enjoys a duality in the secondary and tertiary sectors which does not doom the more autonomous, inward-turning segments to chronic low wages, high failure rates, employment insecurity, and a long-term decline into irreversible disorganization. They would see the fate of efforts at Black capitalism as predictive of the future for whatever efflorescence of Hispanic capitalism might accompany the demographic and political build-up of Mexican-American communities in so many urban centers of the Southwest. If this should prove correct, both the distinctiveness and the vitality of those communities would be diminished, and the symbolic and political significance of these cities for Hispanics elsewhere in the nation would be undercut. But the comparison with Black ghettoes may be as misleading as comparisons with earlier White ethnic enclaves. The situation unfolding in the Southwest is unique in the experience of this country, and neither established models nor past patterns are likely to predict its outcome. Figure 6.1 summarizes this perspective.

THE FUTURE U.S. HISPANIC URBAN ORDER: THEMES AND VARIATIONS

From an airliner four miles up the cities of the Southwest will all seem alike, except for differences in size of sprawl and that build-up of central and outlying nodes that will vary with the total population and expanse of the metropolitan region. But an approach from a lower altitude will reveal some important differences to the discerning

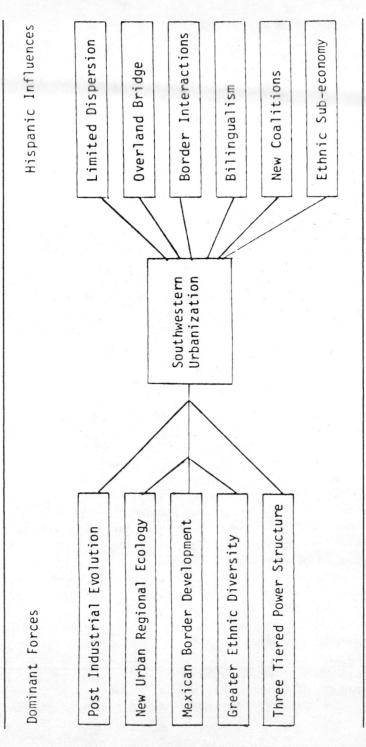

Dominant Forces

Hispanic Influences

Post Industrial Evolution

New Urban Regional Ecology

Mexican Border Development

Greater Ethnic Diversity

Three Tiered Power Structure

Southwestern Urbanization

Limited Dispersion

Overland Bridge

Border Interactions

Bilingualism

New Coalitions

Ethnic Sub-economy

FIGURE 6.1   The Hispanic-American Urban Order

observer. There will be fewer high-rise structures, and they will in general be smaller in scale in the border cities' central areas than in cities of similar size elsewhere; the build-up in *las ciudades de hispanos* north of the border will not be as great as in the Anglo cities, while the external nodes around the Hispanic cities will be relatively greater than those of the outer metropolis elsewhere. This physical structure embodies the flow of Anglo capital into business and gentrification developments.

THE HISPANIC-AMERICAN SOUTHWEST: A SCENARIO

It is easy for the imaginative but fearful to perceive the combination of factors just cited as conducive to Hispanic nationalism and eventually even secessionism in the U.S. Southwest in the next half century (Downes, 1977). A much more probable future is less violent, but still anticipates major and stressful changes for the United States. Culturally and linguistically the border region between the United States and Mexico will contain the United States' most pluralistic and yet also most ethnically segregated cities, and the tensions within the multilingual polity will be intense, the coalitions unstable.

CROSS-BORDER INSTITUTIONS

Three or four decades from now the twin cities will have many more gateways through which they can open into one another. In their efforts to cope with their common problems, these binational urban areas will produce a number of ad hoc, unstandardized arrangements. Informal and formal linkages in the public sector will include water and sewage systems, health, environmental protection, social services, public works, transportation, and planning. A special binational police patrol will be in operation, and although the criminal justice system will not be well integrated, cable television will facilitate collaboration between judicial authorities in prosecutions. In some twin city areas very limited binational metropolitan agencies will be established as a logical extension and further formalization of well-developed linkages. In the private sector binational enterprises will appear in merchandising and all business related to tourism, and then in areas related to urban development and local and regional investments.

This process does not lead to the merger of the municipalities or the disappearance of distinctive identities on the part of most residents. But even the limited extent to which twin cities generate a network of binational agencies and activities, and to some degree seek to align their policies and practices, will introduce into both Mexican and U.S. national affairs problematic influences and challenges to regional policies.

PREDISPOSITIONS AND POLICIES

In central cities where Hispanics are predominant, cultural predispositions are likely to be translated into civic policies in the areas of public education, social service and assistance, and political organization. These policies will influence other institutional systems, including law enforcement, religion, public recreation, and the mass media.

Public Education

In the predominantly Anglo, well-to-do areas of the outer metropolis private schools, as well as at-home and neighborhood satellites using home computer and cable technologies, will become major forms of instruction. But for the great majority of central city Hispanics, Blacks, Asians, and lower-income whites, public schools will continue as the main source of education. This segment of the population will be looking for schools that teach basic skills and knowledge useful for employment or further education, that are orderly and safe, and that are cost-effective. But for practical as well as ideological reasons, Hispanics will continue to push for schools where Spanish is an in-use language along with English. Tensions will develop among minority groups over priorities and agendas, but none of them are likely to be willing to spend scarce funds for desegregation involving mainly each other.

Social Service and Assistance

The majority of Hispanics view the welfare bureaucracy with suspicion, and prefer to survive through the assistance of the extended family whenever possible. As they gain power in the central cities, they will require agency changes to meet their ethnic cultural priorities. In the Hispanic central cities the system will be integrated with the supportive efforts of the family and community. It will utilize modest neighborhood clinics/community centers to connect families with professional services and sources of aid. Expenditures enabling kinship groups to care for children, the elderly, and other dependents will be preferred over institutional settings. The workers employed by these networks will be from within the community.

Political Organization

The long-term decline in municipal power, along with current fiscal retrenchment and the lean years ahead, is largely compatible with the governance needs and political interests of the relatively affluent majority of the Anglo outer metropolis. Over the next few decades private enterprise will provide the majority of services for this segment, and the municipalities there will serve as vehicles for planning and zoning functions oriented to maintaining real estate values, mercantile and recreational amenities, and rituals of local identity. Party distinctions will be irrelevant at the local level, but at state and national levels the citizenry will orient to party loyalties. For officials at this level the claims and commitments generated by the central city municipality will contrast sharply with the posture of largely self-sufficient outer metropolis communities. They will seek to deflect or defeat these claims by their continuing economic and political influence.

Party affiliations will be of minor interest to most Hispanics who move into political roles during the eighties and nineties. For them, political leadership, elected office, and public sector employment generally will represent career opportunities, along with some collective capability to shape community law and policies, to allocate very scarce resources, and to bargain with state and national power groups. Hispanics will operate locally and statewide primarily through groupings reflecting barrio identifications and coalitions, class interests, and to some extent ideological differences about relationships with both the larger U.S. society and Mexico. For most Hispanic residents politics will remain local. Hispanic cities will struggle with the issues of representation and control under inevitably manipulative conditions.

DREAMS THAT SEE BEYOND THE YEARS

As the twenty-first century swings into its second quarter, the cities of the United States will be neither gleaming alabaster nor "undimmed by human tears." But a period of transformation now markedly underway in the Southwest will be largely complete. The turbulence and episodes of urban violence of the eighties and early nineties will be fading memories, like those of the sixties are today, and will be part of the generation gap between those who participated in the social movements prior to the turn of the century and their teenage and young adult offspring. Even the bitter arguments over the post-2000 Border Areas Treaty concluded between Mexico and the

United States will have dropped from the political charts, however loudly and uselessly reiterated by the embittered rear-guard remnants of Anglo supremacy. This does not mean that the United States after 2025 will have entered a near utopian period of ethnic pluralism even in the heavily Mexican-American Southwest. But the accommodations hammered out in the binational, multicultural arenas of the twin cities will have become models of Hispanic-Anglo relations elsewhere in the country. While only a small percentage of the residents of the still-growing binational metropolitan areas will have opted for dual citizenship, far more will be both bilingual and bicultural, and consider themselves Meximericans or Hispanglos regardless of legal nationality. Whatever the changing success and failure rates of the political, economic, educational, and human service innovations undertaken in the region, it will have a liveliness that will both stimulate and irritate the national societies to the north and south.

Is this glance around the corner of the century's turning too optimistic? If Mexico's development is even more difficult and delayed than anticipated, mirgratory pressures will persist, maintaining border region instability. Residual Anglo racism might remain more intransigent than foreseen and be acted out more effectively. And, of course, should any superpower find first strike strategy irresistible, then all these contemplations of possibilities will be irrelevant. The anguished survivors in the remnant communities will have little concern for where boundaries are, or for the maintenance of cultural heritages.

NOTE

1. Although the Southwest is usually considered to include Colorado, Texas, New Mexico, Arizona, and California, the focus here is on the last four, all of which have common borders with states in Mexico. Also, the emphasis is on Mexican Americans, the predominant but not sole Hispanic population in these states. It should be remembered that there are major cultural differences between Hispanics from Mexico, Cuba, Puerto Rico, and other Latin American nations.

REFERENCES

ALLMAN, T. D. (1978) "The urban crisis leaves town." Harper's 257 (December): 41-56.
ALVAREZ, R. (1971) "The unique psycho-historical experience of the Mexican people." Social Science Quarterly 52 (June): 15-29.

APPLEYARD, K. and D. LYNCH (1974) "Temporary Paradise?" San Diego, CA: Fronteras.

AVERITT, R. T. (1968) The Dual Economy: The Dynamics of American Industry Structure. New York: Horton.

BEAN, F. D. and B. S. BRADSHAW (1977) "Mexican American Fertility," pp. 101-130 in C. H. Teller et al. (eds.) Cuantos Somos: A Demographic Study of the Mexican American Population. Austin: Center for Mexican-American Studies, University of Texas at Austin.

BECK, E. M., P. M. HORAN, and C. M. TOLBERT (1978) "Stratification in a dual economy: a sectoral model of earnings determination." American Sociology Review 43 (October): 704-720.

BELL, D. (1976) The Coming of Post-Industrial Society: A Venture in Social Forecasting. New York: Basic Books.

BERRY, J.L.B. and J. D. KASARDA (1977) Contemporary Urban Ecology. New York: Macmillan.

BLOOMBERG, W. (1979) "Anglo retrenchment and Hispanic renaissance: a view from the Southwest," pp. 211-244 in J. P. Blair and D. Nachmias (eds.) Fiscal Retrenchment and Urban Policy. Beverly Hills, CA: Sage.

BRACKEN, P. (1980) "Arizona: precursor of post-industrial America." The Futurist 14 (February): 35-48.

BUSTAMANTE, J. A. (1978) "Commodity-migrants: structural analysis of Mexican immigration to the United States," pp. 183-203 in S. R. Ross (ed.) Views Across the Border: The United States and Mexico. Albuquerque, NM: University of New Mexico Press.

CARDENAS, G. (1977) "Mexican labor: a view to conceptualizing the effects of migration, immigration and the Chicano population in the United States," pp. 159-181 in C. H. Teller et al. (eds.) Cuantos Somos: A Demographic Study of the Mexican American Population. Austin: Center for Mexican-American Studies, University of Texas, Austin.

CETRON, M. (1973) An Analysis of the Impact of Advanced Telecommunications Tech ology on the American City. Washington, DC: Forecasting International.

COALE, A. J. (1978) "Population and economic development. The caseof Mexico." Foreign Affairs 56 (January): 415-429.

COLENUTT, R. J. (1972) "Do alternatives exist for central cities?" in H. M. Rose (ed.) Geography of the Ghetto:. Perceptions, Problems and Alternatives. DeKalb, IL: Northern Illinois University Press.

Consejo Nacional de Población (1980) México demográfico. Mexico City: author.

CORNELIUS, W. A. (1981)"Mexico, San Diego and California: Facing 2000 A.D." San Diego: California 2000 Project.

——— (1978) Mexican Migration to the United States: Causes, Consquences and U.S. Responses. Cambridge, MA: Center for International Studies, M.I.T.

COTERO, M. and L. HUFFORD [Eds.] (1980) Bridging Two Cultures: Readings in Bilingual/Bicultural Education. Austin, TX: National Education Laboratory.

DE LA GARZA, R. (1981a) "Demythologizing Chicano-Mexican relations, pp. 88-95 in S. K. Purcell (ed.) Mexico-United States Relations. New York: Academy of Political Science.

——— (1981b) "Public policy priorities of Chicano policital elite." Presented to the Working Group on U.S.-Mexican Border AreaIssues of the Overseas Development Council, Tijuana, Mexico, December 3-5.

DODE, C. (1979) "The next ten years in education." Presented at the Annual Conference of the World Future Society, Minneapolis, MN, October 18-21.

DOMHOFF, G. W. (1971) The Higher Circles. New York: Vintage.

——— (1967) Who Rules America? Englewood Cliffs, NJ: Prentice-Hall.

DOWNES, R. (1977) "The future consequences of illegal immigration." The Futurist 11 (April): 125-127.

DOWNS, A. (1976) Urban Problems and Prospects. Chicago: Rand McNally.

EDMUNDS, S. (1980) "Three economic scenarios for the '80s." The Futurist (December): 12-21.

ESTRADA, L., J. HERNANDEZ, and D. ALVIREZ (1977) "Using census data to study the Spanish heritage population of the United States," pp. 13-59 in C. H. Teller et al. (eds.) Cuantos Somos: A Demographic Study of the Mexican American Population. Austin, Texas: Center for Mexican American Studies, University of Texas at Austin.

FERNANDEZ, R. A. (1977) The United States-Mexican Border: A Political-Economic Profile. Notre Dame, IN: Notre Dame University Press.

FISHMAN, J. (1966) Language Loyalty in the United States. The Hague: Mouton.

GAPPERT, G. (1979) "Employment policy in postaffluent America," pp. 159-181 in J. P. Blair and D. Nachmias (eds.) Fiscal Retrenchment and Urban Policy. Beverly Hills, CA: Sage Publications.

GLAZER, N. (1954) "America's ethnic pattern." Perspectives USA 9 (Autumn): 137-152.

——— and D. MOYNIHAN (1963) Beyond the Meltin Pot. Cambridge: M.I.T. Press.

GOMEZ, D. F. (1973) Strangers in Our Own Land. Boston: Beacon.

HENDERSON, H. (1978) "The emerging countereconomy," pp. 515-530 in J. Fowles (ed.) Handbook of Futures Research. Westport, CT: Greenwood.

HUSS, J. D. and M. J. WIRKEN (1977) "Illegal immigration: the hidden population bomb." The Futurist, 11 (April): 114-120.

KAHN, H. and J. B. PHELPS (1979) "The economic present and future: chartbook for the decades ahead." The Futurist 13 (June): 202-222.

KEEFE, S. E. (1979) "Urbanization, acculturation and extended family ties: Mexican Americans in cities." American Ethnologist 6: 349-365.

KEEFE, S. E., A. M. PADILLA, and M. L. CARLOS (1979) "The Mexican-American extended family as an emotional support system." Human Organization 38: 144:152.

KNOWLTON, C. s. (1975) "The neglected chapters in Mexican-American history," pp. 19-59 in G. Tylor (ed.) Mexican-Americans Tomorrow. Albuquerque, NM: University of New Mexico Press.

LEIBOWITZ, A. H. (1971) Educational Policy and Political Acceptance: The Imposition of English as the Language of Instruction in American Schools. Washington, DC: CAL/ERIC, ED 047 321.

MARTINEZ-GARCIA, G. (1981a) "La política demográfica mexicana y el mercado de trabajo de América del Norte." Mexico City: CENIET.

——— (1981b) "El mercado de trabajo y las interrelaciones económicas entre México, los Estados Unidos y el Canada." Presented to the Working Group on U.S.-Mexican Border Area Issues of the Overseas Development Council and the Centro de Estudios Fronterizos del Norte de México, Tijuana, Mexico, December 3-5.

MARTINEZ-SANDOVAL, R. (1981) "Mexico's population dynamics: a general survey." San Diego State University. (mimeo).

MOLITOR, G. (1981) "The information society: the path to post-industrial growth." The Futurist 15 (April): 23-30.

NEWCOMBE, P. J. (1979) "Some influences on teachers and teaching in 2030: projections, predictions, and a scenario." Presented at the Annual Meeting of the Speech Communication Association, San Antonio, Texas, November 10-13.

NOSTRAND, R. L. (1975) " 'Mexican American' and 'Chicano': emerging terms for a people coming of age," pp. 143-160 in N. Hundley, Jr. (ed.) The Chicano. Santa Barbara, CA: Clio.

PORTES, A. (1977) "Labor functions of illegal aliens." Society (September/October): 31-37.

ROSE, H. M. (1981) "The diminishing urban promise: economic retrenchment, social policy, and race," pp. 183-209 in J. P. Blair and D. Nachmias (eds.) Fiscal Retrenchment and urban Policy. Beverly Hills, CA: Sage.

RUBIO, G. (1980) "Hispanic impact on the political process." Agenda: A Journal of Hispanic Issues 10 (January/February): 8-10.

SAMORA, J. (1975) "Mexican immigration," pp. 60-80 in G. Tylor (ed.) Mexican-Americans Tomorrow. Albuquerque, NM: University of New Mexico Press.

TORRES, A. (1979) "Interview: Art Torres Speaks on the Political Future of Chicanos." Somos 2 (April): 34-37.

U.S. Bureau of the Census (1980) Final Population and Housing Unit Counts. 1980 Census of Population and Housing, Advance Reports PHC80-V-45, Texas. Washington, DC: Government Printing Office.

——— (1979) Persons of Spanish Origin in the United States. Current Population Reports, C3.186, P20/354, March. Washington, DC: Government Printing Office.

7

The Future of Black Ghettos

HAROLD M. ROSE

☐ IN RESPONSE TO MYRIAD FORCES the American city is undergoing significant change in all facets of its existence. In fact, the character and quality of this change are such that the city of the twenty-first century might bear little resemblance to the city of the mid-twentieth century, at least in terms of modal spatial form. Much attention has been focused on the evolving form of the national urban system, with only limited attention devoted to selected components of that form. Nevertheless, Berry (1980) recently made observations about the future of inner cities, and the conclusions reached were very bleak. This chapter will single out yet another component of urban structure for the purpose of stipulating its most likely future during the remainder of the twentieth century. The target of this discussion will be the nation's black ghettos.

After the mid 1960s ghettos received extensive attention for a brief period as a result of a series of violent outbursts that caused both extensive property damage and a wasteful loss of life. During the seventies changes occurring on both a national and international scale—and the attendant problems associated with them—relegated interest in the well-being of the nation's black communities to a lesser position in the scheme of things.

In part, the above position seemed to stem from the view that the collapse of core cities as desired residential environments was partially attributable to the expanding spatial scale of ghetto communities across the country, a position recently expressed by Lowry (1980). In his view the volume of post-World War II black and Hispanic migration to larger urban centers was more than the cities could absorb. This, in Lowry's opinion, has led to increased levels of social deviance and consequently the evolution of zones of desolation within these cities, which severely influenced their chance for recovery.

One might ask if the fate of one segment of metropolitan America, the core city, depends on the fate of the nation's larger ghetto communities. Or, is the future of the nation's ghettos inextricably tied to future patterns of urban development? The author will attempt to grapple with these two basic questions in assessing the urban future of black communities throughout the nation.

RECENT GROWTH CHARACTERISTICS

Before beginning to project, speculate, or conjecture regarding the future of the nation's ghettos, an attempt will be made to provide limited baseline information specifying their current status. But even this attempt will be very general. Nevertheless, it is well known that urban black populations occupy residential areas where they constitute the majority population. From 1940 to 1970 there was only a minimal reduction in the intensity of residential concentration, but during the same period the physical scale of these residential concentrations rapidly expanded (see Figure 7.1). The outcome of this residential assignment pattern has been the creation of extensive single-race communities stratified by social class. The basis for the existence of such communities is generally attributed to (1) discrimination and (2) residential choice.

Arguments abound regarding which of the above constitutes the more critical determinant. But such arguments have little meaning if extended back in time for more than one or two decades. Therefore, the role of choice on residential preference could only be expected to have played a somewhat meaningful role upon the evolution of black residential patterns since 1970. Previous restrictions on choice have led to the evolution of the ghetto configurations that served, in most instances, as modal residential communities for the nation's metropolitan black population. Black community expansion occurred as a result of racial residential turnover, especially in the North and West. The central issue becomes whether this will continue to be the modal pattern throughout the remainder of this century or whether major modifications will be introduced that will alter both the territorial character of black residential space and the quality of life available to this population.

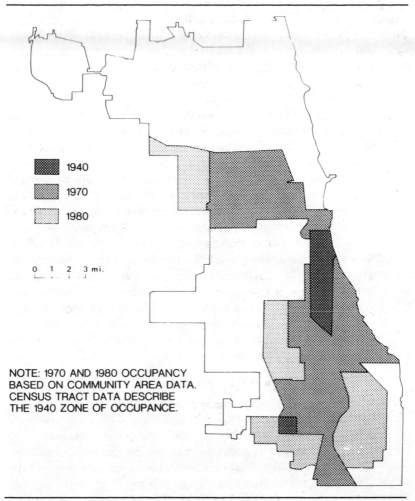

1940

1970

1980

0   1   2   3 mi.

NOTE: 1970 AND 1980 OCCUPANCY
BASED ON COMMUNITY AREA DATA.
CENSUS TRACT DATA DESCRIBE
THE 1940 ZONE OF OCCUPANCE.

FIGURE 7.1   The State of Chicago's Ghetto in Three Time Periods: 1940, 1970,
and 1980

THE GHETTO FORMATION PROCESS

By 1970 the ghetto formation process, which represented the
established way to allocate housing to urban blacks, had been underway for two full generations.

Since the conversion process is now thought to be well-advanced, the term ghetto will be employed to define all spatially concentrated urban black communities that satisfy a critical population threshold (>25,000 blacks) and that are located within a metropolitan context. Since black populations satisfying our critical threshold are present in most of the nation's larger metropolitan areas, we will devote our attention to only a limited class of ghetto centers. Of the nation's principal Standard Metropolitan Statistical Areas (SMSAs) (those with populations of a half million or more), only eleven fail to qualify as ghetto centers. Most often the latter represent new growth centers to which blacks have not yet been attracted.

Two larger ghetto center classes will receive the bulk of our attention in an attempt to project black urban futures. These are (1) principal ghetto centers where the central city black population of the SMSA equaled or exceeded one-half million (super ghettos), and those (2) where population equaled or exceeded 250,000, but was less than 500,000 (giant ghettos). These examples are thought to provide greater insight into the ghetto development process because they best enable us to follow the process from central city to suburban environments. The nation's larger ghettos are associated with these two hierarchical classes. Furthermore, in these SMSAs the black suburbanization process is most advanced, at least in terms of the absolute size of the black suburban ring population (Rose, 1976).

THE STATUS OF PRINCIPAL GHETTO CENTERS

In 1980 the black population in the thirteen ghetto centers satisfying the above criteria numbered 7.6 million, or approximately one-third of the nation's black population. But when the suburban ring populations of these metropolitan areas, which include 2.3 million black residents, are added to the ghetto center totals, the proportion of the total black population is elevated to 38%. Cities that meet the foregoing criteria are identified in Table 7.1. The data shown below represent the population housed in the central city only, therefore constituting a sizable future metropolitan growth pool. Several of these centers already have suburban ring populations exceeding 100,000 persons, implicitly indicating a pattern of metropolitan ghettoization that accelerated during the middle sixties and continues apace. In Los Angeles and Washington the suburban black population is rapidly approaching the size of the central city black population (see Table 7.2). Within these ghetto centers the process of metropolitan ghettoization, as described by Deskins (1981), is most advanced. It is

Table 7.1
The Nation's Principal Ghetto Centers: 1980

Core City	Size of Black Population	Core City	Size of Black Population
New York	1,784,124	Baltimore	431,151
Chicago	1,197,000	New Orleans	308,136
Detroit	758,939	Memphis	307,702
Philadelphia	638,878	Atlanta	282,912
Los Angeles	505,208	Dallas	265,594
Washington	448,229	Cleveland	251,347
Houston	440,257		

SOURCE: *Data User News* (October) 1981: 9.

clear, however, that the pattern is less well developed in southwestern centers.

In each of the above centers, with the exception of Houston, Dallas, and Memphis, the period of traditional core city ghetto development seems to have peaked, with future growth being totally dependent upon natural increase. During the previous decade, net outmigration characterized most of these centers; and in Washington, Cleveland, and Philadelphia the 1980 black population was smaller than in 1970. The most serious problem of net outmigration, however, occurred in St. Louis, where the black population declined by 17.8%. That decline was sufficiently adequate to remove St. Louis from the group of higher order ghetto centers, as it no longer satisfied the population threshold for membership in this group.

One would logically conclude that a thinning out and deconcentration are underway in all first-generation ghetto centers (those satisfying the ghetto center population threshold prior to World War I) and in most second-generation ghetto centers. It seems that only in the newer ghetto centers the process has not yet run its course, and externally induced growth continues to be the primary source of population increase.

For much of this century central city ghettos have occupied varying amounts of acreage in the nation's larger and older cities. At their peak they represented communities of hope that provided an alternative to the level of depressed opportunity prevailing in rural settings and small southern centers. But, as economic change reduces both the role and importance of large central cities in the economy of the near future, the future of some of these communities is now in doubt. Is what we are seeing a simple adjustment to changes in central city

Table 7.2
The Black Population in the Suburban Ring of Principal Ghetto Centers

Center/Rings	Size of Black Population	Center/Rings	Size of Black Population
Los Angeles	438,801	Detroit	128,033
Washington	388,389	Cleveland	84,285
New York	285,315	Houston	79,944
Philadelphia	245,527	New Orleans	79,257
Chicago	230,827	Dallas	50,000
Atlanta	179,220	Memphis	37,848

SOURCE: 1980 Census of Population and Housing, Advance Reports, Series PHC 80.

economies? Or does this indicate the ghettoization process has simply run its course and, like previous ethnic enclaves, the central city ghetto is fast becoming a residual community?

These questions cannot be easily answered with confidence. Nevertheless, there are some tentative answers. On the matter of ghettoization process having run its course, there is a high level of skepticism. Continuing evidence of territorial growth by way of accretion illustrates that most blacks continue to acquire housing by participating in border markets (Berry, 1976). Likewise, evidence illustrating realtor practices designed to maintain existing patterns is still commonplace (Wilson, 1979; Bourne, 1981). Yet the role of the socioeconomic mix of the center's black population should not be overlooked in its influence on the directional bias associated with the spatial pattern of ghetto development. Socioeconomic mix, when affiliated with a population of a given size, is likely to determine if the black population will be essentially confined to a single sector of the city or to multiple sectors.

THE ROLE OF CHANGING CENTRAL CITY ECONOMICS

It is unclear at this point what specific impact changing central city economies and the concomitant residential or housing market adjustments will have on the ghettoization process. As core cities become increasingly oriented away from manufacturing activity and intensify their pursuit of tertiary growth industries, new markets will likely be created that will lead to dissolution of peripheral segments of ghetto space located near downtown. As this process advances and as the pattern of household formation continues to undergo change, the geographic position of ghetto space in the larger ghetto centers is likely to be altered. Unless there are substantial changes in the way housing is allocated to blacks, that alteration would push the ghetto configuration (1) farther from downtown and (2) accelerate settlement in older suburbs.

SOURCES OF BLACK HOUSING DEMAND

In America's larger urban complexes, the pattern of black residential development is clearly influenced by both external and internal factors that interact to produce the resulting outcome. Therefore, one must consider the economic health of the home building industry as a measure of its propensity to satisfy shelter demands of a population in the throes of slow growth, but with household formation vastly exceeding population growth rates. During the seventies the nation's population grew by only 11.4%, while households grew by 27%. Even with high interest rates and rapidly rising housing prices, the industry managed to satisfy a significant element of housing demand brought on by the changing demographic structure of the population, as well as by the changing household formation patterns. Nevertheless, in response to inflationary pressures, a slowdown in residential construction is underway.

How are blacks likely to fare in this new era of housing development, at least as it relates to altering both the form and character of evolving ghetto space? Both the changing demographic and socioeconomic mix of the black population will be significant in determining the level and character of housing demand.

Under what circumstances and where is housing demand likely to be satisfied? In the past most of that demand has been met within the context of an evolving central city configuration. During the eighties, at least within our principal ghetto centers, it appears much of the additional demand will be directed toward selected suburban ring communities. Only among southwestern metropolitan areas will this not represent the principal target zone.

If the evolving spatial pattern of black residential development is not significantly altered, fourth-generation ghetto centers will essentially be confined to a selected set of suburban ring communities located in metropolitan areas where the central city black population already numbers more than one-quarter million or at least did so in 1970.

THE ROLE OF DEMOGRAPHIC STRUCTURE UPON DEMAND

The black population's demographic future will greatly influence housing needs. Providing that no significant changes occur in the pattern of household formation, a slowdown in the rate of natural increase and the subsequent aging of the population should increase demand for single family units.

Although a continued general slowdown in aggregate growth is anticipated, potential for ghetto or nonghetto development will vary as a function of growth within specific metropolitan contexts. Thus, in the short run changing migration patterns are likely to have a more important effect on housing demand than are changing fertility patterns. When growth rates within two urban settings are synonymous, the setting where migration is the principal contributor to growth will generally require a larger stock of housing to accommodate that growth.

Since migration is likely to exert the most pervasive impact upon future housing requirements, the volume and direction of migration appears to represent the key to future ghetto development potential. Declining attractiveness of the principal northeastern and north central ghetto centers should at least reduce the pressure on central city housing stock. Also, the mature population in these cities who have experienced significant income gains over time will increasingly seek to satisfy their altered housing needs in the suburban rings of these metropolitan communities. The magnitude of this suburban demand will be influenced by the size of the upwardly mobile population and by the availability of housing within this population's financial capacity.

The slowdown in south to north migration to numerous principal ghetto centers should lead both to decreased ghetto density and to ghetto decentralization. One writer has attributed the greater taste for residential segregation on the part of blacks of southern origin to have led to a more intense concentration than might otherwise exist (Roof, 1980). This point, however, has not been extensively researched, but were it shown to be valid some additional deconcentration might be expected in numerous principal ghetto centers. If the inverse is true and there is an increase in the volume of northern-born black migrants to the South's principal ghetto centers, they too should show some incremental level of deconcentration.

In the past most black migrants were lower-income persons whose choice of migration destination was based on a network of kinship and social ties linking place of origin with place of destination. That network facilitated the rise of primary ghetto clusters in the nation's ghetto centers. But as populations in these centers grew over time, additional housing demand based directly on the magnitude of present migration flows has assumed a secondary importance. This situation creates two distinct possibilities: (1) that the location of housing demand created by migrants and nonmigrants will differ as a result of the operation of a regional effect; and (2) that centers which were earlier targets of black inmigration will become the primary dispensers of outmigrants whose housing choice behavior may not be constrained by social and kin networks.

Although the population's demographic structure is likely to be significant in establishing housing requirements, the extent to which those requirements are satisfied will be mediated through the population's socioeconomic structure. Thus the social class structure and its attendant changes throughout the remainder of this century will do much to influence the spatial pattern of black residential occupancy.

For our purpose class might best be defined in income terms. On the basis of 1979 black family income distribution, it is possible to establish a series of income groupings to generally determine the potential of blacks to participate in housing markets remote from the ghetto's interior. At that date, according to the Census Bureau, almost one-fifth of the nation's black families had earnings of $5,000 per year or less; 31.6% earned between $5,000 and $15,000 per year; 22.0% earned between $15,000 and $25,000 per year; and only 17.1% earned more than $25,000 per year.

On the basis of the above income distribution, one would conjecture that the general participation level in nonghetto markets is unlikely to exceed one-fifth of the total in the owner occupancy market. Thus, although black income levels have shown a marked increase, that increase has not been great enough for blacks to have access to a broad range of metropolitan housing. But even under these income constraints, a larger share of the incremental demand can be satisfied beyond the shadow of the ghetto than has been true in the past.

The income distribution described above can be loosely interpreted to represent the basic social class mix, with the two lower earning-ability groups representing a lower-status population and an underclass, and the two higher-status population groups representing the working and middle classes. Segments of the latter two strata will most likely choose to participate in nonborder markets; but extent of participation will largely be influenced by barriers to access, including both social and institutional responses.

THE ROLE OF HOUSING SUPPLIER BEHAVIOR

Changes in the black population's demographic structure and economic mix and alterations in household formation patterns acting conjointly have influenced the nature of black housing demand. But the provisions for satisfying that demand rest largely with suppliers. Financial institutions, developers, and realtors, as well as the effectiveness of government policy designed to promote housing access, will be crucial in determining what proportion of the total black housing requirement will be satisfied within or beyond the context of the ghetto. Open housing legislation and other public action have had some effect on lowering access barriers simply by placing the real state industry on notice that legal action will be initiated if past racial practices are continued. Berry (1979) contends that the formal open-housing mechanism in Chicago has been less successful than anticipated because of black lack of interest in integrated housing. This, however, simply points up the complexity of the problem and the conclusions that one can arrive at when employing simple dichotomous categories.

Any attempt to assess the future likelihood that blacks will continue to acquire housing in the principal ghetto centers in a manner consistent with that prevailing for more than two generations must look closely at the changing nature of black housing demand and supplier willingness to act upon that changing nature. Palm (1981) recently illustrated how changes in lifestyle orientation affect housing demand and its subsequent locational consequences in fostering segmented housing submarkets within a metropolitan context.

Historically, however, black communities have tended to be more heterogeneous as they were a product of a dual housing market where segmentation was largely absent. But as these communities attained a critical population mass, segmentation within a dual market became possible. The principal ghetto centers now include black communities that transcend in size the population of many of the nation's rapidly growing small- to intermediate-sized metropolitan areas.

At this juncture and under the present set of rapidly changing circumstances, we need to ascertain whether, in the near future, the segmented housing demand of an increasingly diverse black population is likely to be satisfied within the framework of a dual, "quasi-dual," or open housing market. Supplier behavior will play a significant role in the outcome. Ghetto futures can be realistically addressed only within this context. A logical departure point might be to view the status of present growth and direction of black suburbanization, since it represents the most recent state in the evolution of black residential patterns.

BLACK SUBURBANIZATION AND
GHETTO FUTURES

Although rapid suburbanization has been underway since the fifties, black participation in suburban development did not get underway until the latter half of the sixties (Rose, 1976; Lake, 1981; Goodwin, 1979).

Although before 1970 there was no substantial black presence in the suburban rings of the nation's larger metropolitan areas, even then blacks were confined to a small number of suburban communities and to specific neighborhoods within those communities. Thus, during the period of incipient black suburbanization, racial residential turnover in contiguous space represented the primary mechanisms providing access to housing in this environmental context. The spillover process represented the modal one facilitating black suburban entry, but at the same time it promoted ghetto expansion across political jurisdictions.

If this process continues unabated, then blacks will simply be required to express a housing preference within the confines of a channelized market that is itself simply dependent upon the magnitude of black demand and the subsequent ability of whites to have their housing demand satisfied in alternate channels. By observing how black suburban development has proceeded in a single principal ghetto center, it is possible to loosely examine the extent to which there has been a deviation in the process.

THE BLACK SUBURBAN EXPERIENCE:
THE CASE OF CHICAGO

Chicago represents one of the nation's larger and older ghetto centers, and until the most recent decade it was the recipient of a large and continuous flow of inmigrants from the central city and net inmigration to the suburban ring taking place.

Black movement to the suburban ring was slower in Chicago than in numerous other ghetto centers during the post-1965 period. This might be attributed to the failure of the primary ghetto to reach the city's edge prior to 1970, thereby reducing potential for spillover. But continued ghetto growth in each of the city's primary ghettos and competition with recent immigrants to the city have resulted in energizing suburban housing demand. The objective is to specify how that demand has been satisfied and to what extent it complements or diverges from the established housing allocation pattern.

By 1970 10.4% of the black population in the Chicago SMSA resided in the metropolitan ring. Most of these persons, however, resided in a series of old industrial satellites, in upper-income service centers, or in black colonies.

These ring residents seldom participated in an open housing market, but were almost entirely associated with one of several genres of black housing allocation mechanisms. Thus the growth of the suburban ring population during the seventies reflected the incremental demand generated by an inplace black population, pent-up demand inbedded in the core city ghetto, and demand created by persons from outside the SMSA.

CENTRAL CITY SPILLOVER AND SUBURBAN DEVELOPMENT

During the most recent decade there was a continuous push of the black housing market toward the periphery of the city. Seven additional community areas, zones containing generally in excess of 40,000 people, were added to the ghetto total during this period, although three had been heavily penetrated during the sixties. But by 1977 "Chicago seemed poised on the edge of a similar experience as both westside and southside ghettos spilled over the city limits. As this happened, middle-class black Chicagoans had available to them an increasing array of housing opportunities" (Berry, 1979). In Berry's view the three white communities (Calumet Park, Dolton, and Riverdale) situated between the edge of the southside ghetto and that crescent of earlier established black suburban communities would simply become a part of a single ghetto community by 1990.

Approximately 80,000 blacks from elsewhere established residence in the suburban ring during the seventies. Most no doubt moved from Chicago to other parts of Cook County, although there is evidence of limited movement to each of the other ring counties, with the exception of McHenry.

Although there are more than 200 suburban communities in the Chicago metropolitan area, one would expect major differences in level of access. Easiest access should logically occur among the band of communities abutting the edge of the city, with access becoming more difficult with increasing remoteness. Thus middle-income blacks moving to suburbia have been basically confined to housing in the older suburbs, whereas middle-income whites tend to settle in the new suburbs (Stahura, 1982). This pattern simply facilitates ghetto expansion, with the point of the ghetto now wedged in the suburbs rather than in the central city. Although access is expected to diminish with distance from the edge of the city, we do not assume that no black households will be unable or unwilling to seek housing remote from the advancing ghetto. A review of the changes occurring in Chicago in the most recent decade can be employed to substantiate or invalidate the above assumptions.

CHANGING HOUSING DEMAND

If black married couple households have done as well in Chicago as they have nationally in terms of median family income ($21,000 in 1979), one would expect them to be able to participate in various submarkets based on housing costs. Since newer housing is generally more expensive than older units, housing cost should be expected to increase with distance from the city's edge, at least in Cook County. Thus, unless most black incomes are concentrated in a narrow band around the median, we should not expect a piling up of demand in edge communities. The extent to which crowding does occur, however, is a crude measure of access or simply a willingness to participate in border markets. Recently it has been shown that black professionals are willing to participate in nonborder markets in order to acquire those strands in the housing package they value (Rose, 1981). One strand that has become increasingly valuable is that of amenities associated with recently constructed units. In Chicago most newly constructed units have been built some distance from the city's edge.

During the seventies black movement to the Chicago suburban ring was concentrated in a limited number of communities, mostly older declining communities well on the way to becoming essentially black communities, such as Maywood and Harvey. Others were established middle-income communities characterized by status persistence; still others were growth communities where new housing stock increased substantially during the decade. Thus, blacks did indeed enter a range of community types where different strands in the housing package might be thought to have assumed differential importance. Nevertheless, the greatest numerical increase in the black suburban population occurred in ghetto communities or on the periphery of black colonies established during an earlier period.

THE ENTRY COMMUNITIES

Fewer than thirty communities were the primary settlement targets. The largest volume of settlement was confined to a small group of

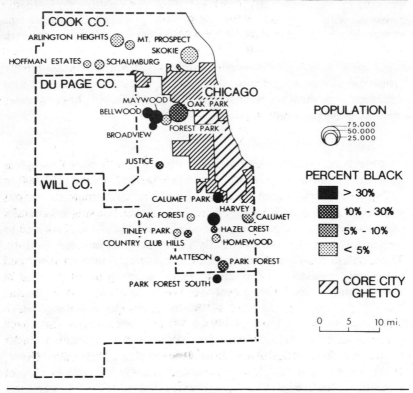

FIGURE 7.2   Chicago's Black Suburban Entry Communities, 1970-1980

southern and near western Cook county communities (see Figure 7.2). A secondary settlement cluster was located in northwestern Cook County in a zone that De Vise (1976) indicated had been the locus of much employment relocation. The volume of anticipated spillover along the edge of the southside ghetto failed to materialize, although some entry was evident. Housing values in these communities suggest that these were not communities that would be attractive to middle-income blacks. The economic characteristics of the white residents would also suggest a lessened likelihood of flight. This combination of factors possibly slowed the spillover process.

The largest growth cluster was a group of communities centered on Harvey and its environs. Thus middle-income blacks were inclined to replace middle-income whites in a group of communities where little new housing was being built.

The second principal growth cluster is in close proximity to Chicago's westside ghetto. Entry has been concentrated in Maywood,

Oak Park, Broadview, and Bellwood. Oak Park, which was less than 10% black in 1970, became a primary growth center in the following decade. Because of white response to perceived impending change, this community's actions have been well publicized in the press. Oak Park has used various integration management techniques designed to disperse blacks throughout the community, rather than simply allowing the wave-like pattern of change that occurred in other communities.

As a means of attempting to reach an equilibrium that would prevent loss of white demand, integration management has been employed elsewhere in the suburban ring. It is uncertain how effective these techniques have been and if they are able to disperse black demand over a larger segment of the housing market. But unlike communities to the south, there is strong evidence of a spillover effect in this westside cluster. Yet that effect seems highly channelized as the level of entry in communities to the north and south appears weak.

CONCLUSIONS

A review of the ghetto development process during the most recent decade provides strong support that ghettos will continue to be the primary housing providers for blacks for the rest of this century and early into the next. There is little doubt that a growing percentage of black housing demand will be satisfied outside of this context, but that share will be inadequate to lead to a major reversal in the ghetto formation process. These conclusions, however, are largely based on a continuation of the internal workings of the housing market and the seemingly stable white response to a black presence at other than nominal levels. Thus, for major changes to occur, the environmental preferences of both whites and blacks would need to undergo a major transformation.

The change currently taking place in the nation's principal ghetto centers represents a response to external forces. In the near future the nation's principal ghetto centers outside the South are likely to become exporters of people, assuming the role performed by the rural South earlier in the century. If this anticipated function evolves, how will it influence the ghetto formation process in secondary ghetto centers? Or will there reemerge regionally specific ghetto formation processes? These are questions for which there are no easy answers at this time.

This chapter has focused principally upon the territorial black community development. A more important feature, however, is what the quality of life will be in the evolving configurations by the end of the century. If the pattern of socioeconomic isolation that seems to be emerging in the principal ghetto centers continues unabated, the potential for internal conflict is enhanced.

Under conditions of severe economic deprivation, wherein the working poor would become the predatory targets of the nonworking poor, jungle-like conditions could emerge in these isolated clusters. On the other hand, low-income residents of the principal ghetto centers might be induced to migrate toward urban growth poles as a means of alleviating stress under conditions of economic deprivation. Needless to say, there appears to be little concern regarding the social and economic implications associated with the present spatial reorganization upon the future of urban blacks, or for that matter upon the future of the city. But one thing is clear, and that is most blacks will reside in residential clusters where they constitute the majority population well into the next century unless unforeseen external forces create alternative futures.

REFERENCES

BERRY, B.J.L. (1980) "Inner City Futures: An American Dilemma Revisited." Transactions of the Institute of British Geographers: 1-30.

— — — (1979) The Open Housing Question: Race and Housing in Chicago, 1966-1976. Cambridge, MA: Ballinger.

— — — (1976) "Ghetto expansion and single-family housing prices: Chicago 1968-1972. Journal of Urban Economics 3: 397-423.

BIANCHI, S. M., R. FARLEY and D. SPAIN (1982) "Racial inequalities in housing: an examination of recent trends." Demography 19 (February): 37-52.

BOURNE, L. S. (1981) The Geography of Housing. New York: V. H. Winston.

DESKINS, D. R., Jr. (1981) "Morphogenesis of a Black ghetto." Urban Geography 2 (April-June): 95-114.

DE VISE, P. (1976) "The suburbanization of jobs and minority employment." Economic Geography 52 (October): 348-362.

GOODWIN, C. (1979) The Oak Park Strategy. Chicago: University of Chicago Press.

LAKE, R. W. (1981) The New Suburbanites: Race and Housing. New Brunswick, NJ: Rutgers University Press.

LOWRY, I. S. (1980) "The dismal future of central cities," in M. Solomon (ed.) The Prospective City. Cambridge, MA: MIT Press.

PALM, R. (1981) The Geography of American Cities. New York: Oxford University Press.

ROOF, W. C. (1980) "Southern birth and racial residential segregation: the case of northern cities." American Journal of Sociology 86: 350-358.

ROSE, H. M. (1981) "The black professional and residential segregation in the American city," in G. Peach, J. Robinson, and Q. R. Smith (eds.) Ethnic Segregation in Cities. London: Crooan-Held.

— — — (1976) Black Suburbanization. Cambridge, MA: Ballinger.

STAHURA, J. M. (1982) "Status transition of Blacks and Whites in American suburbs." Sociological Quarterly 23 (Winter): 79-93.

WILSON, F. D. (1979) Residential Consumption, Economic Opportunity, and Race. New York: Academic Press.

Part III

Forces of Transformation:
Will Cities Enjoy the Future?

□ EXCEPT FOR A FEW COLONIAL EXCEPTIONS, most American cities are relatively new and are unlikely to have experienced more than two or three major transformations of their functions, structures, and forms. Between the forces generated by the post-affluent transition in the early 1970s and those trends and reactions emerging as the current post-affluent decade continues, a sense of substantial transformation in urban development is likely to be widely held as cities approach the end of the twentieth century.

Van Til outlines some of the spatial consequences of four energy scenarios in the near and immediate future and elaborates some of the problems of an energy-short city in the twenty-first century. Coates lays out seven ways in which technology is likely to influence cities in the next fifty years. Embryonic technological development combined with the natural turnover of urban infrastructure should combine to produce substantial shifts in the functional continuity of cities although visual and structural continuities will persist into the twenty-first century.

Hirschhorn reports on the revolt of the middle managers in the year 2001 and reveals some likely transformations in the institutional environments of cities. Saegert reviews issues associated with the gender identity of built environments and speculates about an androgynous city in the future. Blair presents a model of how the expansion of economic dislocations is going to create more urban irregular economies alongside the changing mainstream economy.

These chapters all suggest that in the complex society of the twenty-first century, the cities will be even more complex.

8

New City Types in an Energy-Short World

JON VAN TIL

□ SOCIAL SCIENTISTS HAVE NEGLECTED the study of the future as an area for empirical study, apparently believing that images of the future do not directly affect the behavior of individuals and groups. Considerable literature in the applied behavioral sciences contradicts this judgment. Ronald Lippitt and his colleagues, for example, have found that conceptions of the future greatly affect individuals' willingness to entertain alternative choices of productive action. The ability "to future" is increasingly coming to be seen as a critical aspect of what Amitai Etzioni has called the "active society."

The ability to engage in energy futuring is particularly important because energy policy is likely to be most importantly set at the level of the individual consumer as he/she makes decisions of automobile use, home weatherization, and food production and consumption. Social scientists can advance both knowledge about these choices and the widest selection among them by making belief, behavior, and lifestyles central topics on their research agendas.

Belief—because we start with the understanding of the moment, and build from that base to try to understand the future. Behavior—because our actions are themselves embedded in patterns we inherit from the past, and these actions are the starting point from which our future acts will develop. And lifestyle—because the ways we live gratify or frustrate us in proportion to our expectations, and our images of the future will determine, at least in part, how we will experience that future (Van Til, 1982).

The future is of course unknown. It is also unknowable in the many elements that will come to form it as it unfolds day by day. Despite these uncertainties, it is nonetheless important to try to understand what the future will bring.

In the area of energy, for example, the future might well provide, at least for those of us in the developed world, continuities of supply and availability. As one set of sources of energy (particularly oil and gas) dwindles, other sources would more amply replace them (such as nuclear fusion and solar energy). In this way energy could continue to enable urban expansion and decentralization.

Such a happy energy future is not the only possibility, however. It is not out of the question that, even in the energy-rich United States, we will be able to achieve a painless transition from fossil fuels to the renewable sources of solar and fusion energy. Oil embargoes by suspicious overseas suppliers, interruptions in coal development, continuing difficulties in the nuclear industry, and even the possibility of war itself stand in the way of our achieving energy sufficiency. It is quite possible, if not probable, that our energy future will be a troubled one.

This chapter is about the form our urban society might take if a number of the unfortunate circumstances just listed come into reality. I will ask four questions: (1) Can energy patterns transform our cities? (2) What urban and spatial forms are implied by shortfall? (3) What alternative paths lead from where we are to the energy-short city? (4) what would it be like to live in an energy-short city?

CAN ENERGY PATTERNS
TRANSFORM OUR CITIES?

Not only can energy shape our cities in the future, but energy patterns have fundamentally shaped cities from the beginnings of urban development. The preindustrial city, as Gideon Sjoberg has so clearly demonstrated, was built on a transportation base of horse travel and foot. So were our own eighteenth-century U.S. cities. Where energy and technology cannot sustain mechanized transit, the rich live at the center of urban developments and the poor commute by foot from their suburban shanties to the urban center.

With the development of the horse-drawn streetcar in the mid-1800s and its electrification in the late-nineteenth century, the "walking city" disappeared (Warner, 1962). It was replaced by the metropolis of increasing scale so familiar to late-twentieth century denizens of the Western world. Center city became a place of work and not residence; inner cities became the residential locus of the poor; and the well-off steadily extended the range of their automotive commuting toward the small towns and rural areas beyond the fringes even of suburbia. The streetcar, and most particularly the gas-fueled automobile, facilitated the emergence of contemporary suburban American.

The increasing scale of metropolitan communities reflects the happy history of energy transitions. When the capacity of nearby forests to supply sufficient wood to heat our eighteenth-century homes dwindled, coal became available in sufficient supply to permit the massive industrial development of the nineteenth century. When coal became less accessible, the development of oil allowed a further increase in energy supply sufficient to permit unparalleled personal mobility by automobile.

Now we sit on the cusp of the next energy transformation, awaiting the development of nuclear fusion sufficient to provide for our renewable energy needs until the last generation. Such a transformation, if achieved, would permit us easily to maintain, if not expand, the spread patterns of what Berry (1980) calls our "counterurbanizing" society. Taken together with the rapid institutionalization of the communications revolution, this energy-rich future would permit the renewal of computerized cottage industry, the use of available oil and gas for personal transit only, and the extension of rail transport using nuclear sources. It presages a twenty-first century city that becomes more fully decentralized, its residents spread widely through the countryside that surrounds the modest remnants of the urban core.

Enter into this fulsome image of our energy future the "What if" factor: What if, despite our increasingly guarded hopes for nuclear energy, fusion energy cannot be commercially developed through the twenty-first century? What if, despite our riches in this resource, coal proves increasingly difficult to extract and burn, its industry plagued by strikes, environmental limitations, and concern for pollution? What if, despite our knowledge that it can become a major energy source in the twenty-first century, we continue to construct buildings in nearly total disregard for the gains solar energy might provide? What if, despite our fullest placement of military orders and continuing demand, world realities impel the major foreign suppliers of oil to impose long-term embargoes on American importers? And what if, as the great majority of Americans expect in 1982, a major war erupts, both diverting resources to military use and endangering sources of energy supply?

Such contingencies could, despite the rampant optimism of both governmental and industry leaders, seriously disrupt the American energy economy and lead to a period of prolonged energy shortfall.

They would, like the previous energy landshifts, fundamentally affect the ways in which we would choose our places of work and residence. They would, in short, once more transform the faces and shapes of our cities.

WHAT URBAN AND SPATIAL FORMS ARE IMPLIED BY SHORTFALL?

In the event of a profound and prolonged energy shortfall, forces that would impel both spatial centralization and decentralization would be set in motion. The precise determination of these forces is not easily predicted, but the major directions of change can be clearly foreseen.

The most important centralizing force involved in shortfall would be the drastic reduction in the amount of oil available for the individual automobile owner. Whether occasioned by war or the progressive exhaustion of existing and future reserves, oil is the most likely energy resource to fail us, possibly before the onset of the twenty-first century.

Without easy and abundant oil, a fundamental transformation will be required in patterns of personal transport. Automobiles will be rendered increasingly expensive to operate as gasoline prices rise quickly, and conversion to alternative fossil fuels on a large scale will not be easily achieved.

The anticipated societal response to a decline in personal transportation involves, in the short run, conservation, and adds, for the middle and long term, the development of "compact cities." Clustering residential, service, and industrial facilities not only reduces the need for personal transport, but it also allows energy saving by means of common walls and ceilings, heat islands, wind buffers, and energy cogeneration.

A return to the compact city would not be accomplished in a matter of months, or even years. It would take a prolonged pattern of energy shortfall to encourage the concentration of new and rehabilitated structures, and the altering of patterns of extended commutation. Such a reurbanized pattern would, in all likelihood, not involve the recreation of gigantic megacities with a single urban focus. Rather, it would generate a number of urban nodes within each metropolitan area. Each of these nodes would become a center around which the largest part of the economic, social, and cultural lives of its constituent residents would cluster (see Van Til, 1979; Burton, 1979; Leven, 1978).

In the shorter run the increased salience of conservation that would accompany shortfall would set in motion a number of other forces. Residents would seek to upgrade the solar capacities of their homes; they would turn wherever possible to the growing of some of their own food; and they would seek to maximize their recycling of existing resources. In each case those persons resident in suburban or nonurban communities would be advantaged over urbanites, since solar collection, gardening, and recycling are activities that share a common need for land.

The competing forces of centralization and decentralization are likely to each be sufficiently strong to restrain an all-out move back to the cities or out to the hinterlands when energy shortfall threatens. Rather, these forces seem more likely to impel all residents, whether urban, suburban, or rural, to seek to maximize the energy advantages of their respective communities.

A similar logic leads to the prediction that energy shortfall would not lead to either a rush to or from Sunbelt or Frostbelt, but rather to the selective enhancement of the particular energy advantages of each region. For energy survival in an era of shortfall, then, my hypothesis is that most Americans will cast down their buckets where they are, to paraphrase Booker T. Washington. Or, at least, they will seek to preserve the advantages of their respective niches in the metropolitan ecology without making fundamental changes in community, region, or lifeways. This hypothesis is more fully explored as one of a series of alternative paths from the present to the twenty-first century future.

WHAT ALTERNATIVE PATHS
LEAD TO THE TWENTY-FIRST CENTURY CITY?

Futurists increasingly tend to employ multiple scenarios in seeking to understand emerging futures (Amara, 1981; Cornish, 1979). In a recent book on the impact of energy patterns on urban development, the author (Van Til, 1982) identified four such scenarios (See Figure 8.1).

Scenario I may be seen as the "hard path transcendent," or "Reagan's hope." In this future a successful transition is made from the nonrenewable fossil fuels of the twentieth century to a mixed reliance on coal and nuclear energy in the twenty-first century. This transition involves neither shortfall nor extensive solar development. Nor does it require any substantial attention to energy conservation. It allows the citizens of the United States to continue to enjoy energy plentitude without worrying about the need to reduce the present condition of energy waste.

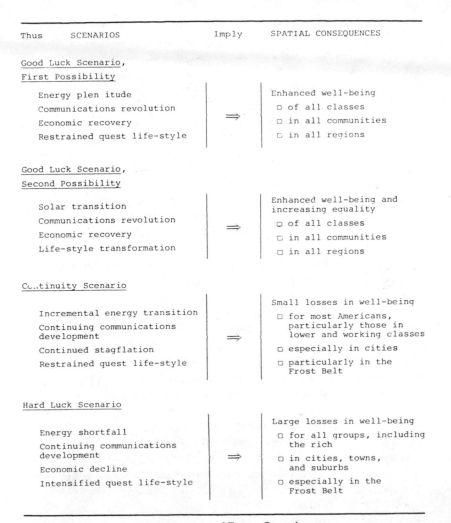

Thus SCENARIOS	Imply	SPATIAL CONSEQUENCES
Good Luck Scenario, First Possibility Energy plenitude Communications revolution Economic recovery Restrained quest life-style	$\Rightarrow$	Enhanced well-being □ of all classes □ in all communities □ in all regions
Good Luck Scenario, Second Possibility Solar transition Communications revolution Economic recovery Life-style transformation	$\Rightarrow$	Enhanced well-being and increasing equality □ of all classes □ in all communities □ in all regions
Continuity Scenario Incremental energy transition Continuing communications development Continued stagflation Restrained quest life-style	$\Rightarrow$	Small losses in well-being □ for most Americans, particularly those in lower and working classes □ especially in cities □ particularly in the Frost Belt
Hard Luck Scenario Energy shortfall Continuing communications development Economic decline Intensified quest life-style	$\Rightarrow$	Large losses in well-being □ for all groups, including the rich □ in cities, towns, and suburbs □ especially in the Frost Belt

FIGURE 8.1   Urban Consequences of Energy Scenarios

SOURCE: Permission of the Westview Press is gratefully acknowledged for the use of Figure 1, which first appeared in Van Til (1982).

Scenario II maybe called the "Solar/Conservation Transition" (see Solar Energy Research Institute [SERI], 1981; Harman, 1979). In this future a successful transition is made from the oil and gas of the twentieth century to a renewable nonnuclear base in the twenty-first century. This transition involves substantially reducing the present cushion of energy waste (Krenz, 1980). It does not, however, involve stringent sacrifices or deprivations. It also protects society from what is often viewed as an unhealthy and dangerous reliance on nuclear technology (See Lovins, 1977; Commoner, 1979; Solomon, 1980).

Scenario III may be called the "muddling through," or "more of the same" scenario. Here no decisive choice is made between the hard and soft paths. Energy policy seeks to encourage the development of all energy forms in the hope that enough of them will prove sufficiently fruitful to provide continuing energy adequacy. Through a combination of market forces and governmental incentives, a "mixed" package of energy development is expected to emerge, while economic development proceeds forward at an uneven pace.

Scenario IV, in contrast to the previous three, envisages a substantial failure in energy development. Here the wrong choices are made—say of nuclear development which proves unreliable and expensive—or of oil which proves quickly exhaustible. Foresight is not employed to achieve an orderly transition to any adequate energy base, and the entire national economy sinks deeper into the disastrous mire of permanent recession and stagflation.

IMAGES OF URBAN FORM IN THE SCENARIOS

Each of the four scenarios involves its own particular urban configuration. Scenario I enables the continuing spread of counterurbanization, and foresees a future in which scale is extended to its maximum. "Persons of greater wealth and leisure will find homes and work among the more remote environments of hills, water, and forest, while most will aspire to this as an ideal (Berry, 1975: 132). The city, in this scenario, becomes the home of the poor and the locus of certain increasingly insignificant economic and cultural activities.

Scenario II involves much less counterurbanization, since it implies that residents in all regions and communities will seek to adapt more effectively to their various energy situations. Conservation gains are expected from urbanites and suburbanites, Frostbelt and Sunbelt residents alike in this future.

Scenario III similarly involves the quest for energy adaptation without either more or less urbanization. Nonetheless, this scenario involves a list toward the Sunbelt, in that economic stagnation is expected to gnaw most painfully at the lower social classes within the Frostbelt states. The Continuity scenario, should it persist into the twenty-first century, foresees a continuing growth in the South and West, unaccompanied by the economic renewal of the northeastern and north central states.

Finally, Scenario IV sees the increasing destruction of cities, particularly in the Frostbelt, as centers of social diversity, economic productivity, and cultural creativity. In one presentation of this scenario: the Frostbelt city becomes the home of an ever-larger population of deprived and desperate individuals and families. These urban ranks are swelled by former members of the working and middle classes who find themselves unwilling draftees in the war on inflation. The few persons of high income who remain in the cities seek to defend their neighborhoods or high-rise towers from the ravages of increasing urban crime and violence. Numerous cities abandon their systems of public transit, as Birmingham did temporarily in 1981. Public school systems begin to close in the Frostbelt cities in the mid-1980s, and organized rebellion begins to emerge in a number of cities.

The suburbs of the Frostbelt become the refuge for many middle-class urbanites who are able to escape the reach of disorder and declining public service. But their housing aspirations exceed their ability to pay, and many suffer real losses in housing quality as they seek to escape the city's reach. The deregulation of natural gas prices in 1985 leads many suburban families to the brink of foreclosure. Their lines of communication also become stretched beyond the comfort of family stability, especially in the increasingly prevalent two-career family.

A strong market develops in the city for returning middle-class families, desperate for inexpensive housing closer to their places of work. Seeking to save on both their energy bills and their mortgage payments, these prospective residents organize collectively to pressure city mayors for adequate services as a condition for their migration. Faced with the opportunity to increase the tax base of the city with employed residents, as well as the threat from commercial and corporate firms to move to the suburbs, large-city mayors turn to policies of expulsion of the poor. One-way bus tickets to Sunbelt cities are provided by welfare agencies to the unemployed and their families; low-income housing that does not meet code standards is demolished, and income maintenance programs discontinued by the federal government are not replaced at the local level. State legislation facilitating the construction of state-funded housing for the unemployed on the edges of metropolitan areas is vigorously supported by big-city mayors, most of whom are black.

Meanwhile, in the Sunbelt a higher level of prosperity is maintained as increasing military expenditures provide for sustained economic growth. Concern rises about the possibility of an influx of unemployed Frostbelt families, and high barriers to migrants are erected in the form of five-year residency requirements for the receipt of any public service save preventive health and public education (Van Til, 1982).

In Scenario IV by the twenty-first century cities like Philadelphia, Buffalo, Detroit, and Chicago will have become vast outposts for the urban underclasses, their affluent center cores ringed by unrelieved scenes of desolation, suffering, and decay. What remains of urban vision and culture is left to the spread cities of the South and West, but they retain their essentially suburban character. These cities serve as the major refuges of affluence in an America in economic and social decline.

THE ROLE OF POLICY

Each of the first three scenarios found a political champion in the 1980 presidential campaign, and each is likely to remain a political standard in the years ahead.

President Reagan is, of course, the supreme defender of the faith in hard-path plenitude. His energy policy aims to establish a nuclear base while rapidly expanding and exploiting remaining oil and gas reserves. His position denies the shortfall possibility, and ridicules the conservation strategy. It is a faith, which if proven unfounded, could commit the United States to a long-term endurance of Scenario IV.

President Carter, on the other hand, served as the preeminent practitioner of the pragmatic politics of Scenario III. His energy policy sought to enhance all options, and to facilitate the emergence of a balanced package of renewable and nonrenewable, nuclear and solar, conserving and yet adequate energy supply. Such pragmatism provided something for everyone, but also aroused the opposition of many. Historians may come to judge the policy more kindly.

Democratic challengers Edward Kennedy and Jerry Brown, as well as minor-party candidate Barry Commoner, spoke as proponents of the Solar/Conservation Transition in 1980. Their campaigns indicated the presence of their position, and kept alive the possibility of Scenario II.

The emergence of energy-based ideologies, first in the environmental movements of the 1970s, and then in the national politics of the 1980s, lent definition to three distinct visions of the future of the United States.

These ideologies of energy, as revealed in the 1980 presidential campaign, are those of energy plentitude on the right (enough oil will be found if the price is right, synfuels can be developed, conservation means deprivation), energy pragmatism in the center (both conservation and selective development should be pursued), and energy transformation on the left (small-scale, appropriate technology solutions and new conserving lifeways are preferable to continued hard energy development). These positions may be arrayed from the optimism and continued faith in progress of the conservative position to the rejection of the belief in continuing technology of the radical left. However, many on the left remain optimistic regarding prospects of developing sufficient renewable energy for the future.

These energy ideologies are largely consistent with the tradition of conservative, liberal, and radical thought, respectively. Conservatives have traditionally voiced their trust in the powers of the free market, backed by a strong national military force and interest. Conservatives have conventionally discounted notions of the perfectibility of man in building the good society, and have also, in recent years, been more than willing to bend their faith in the free market to sustain the economic power of struggling corporations. Overall, a faith in the free market, the corporation, and the nation-state sustains the allegiance of conservatives, and they center their energy policy on those forces in selective combination.

On the left, in contradistinction, the faith has always been strong that individuals will be able to rise to the challenges of social crisis if assisted and supported by wise collective political action.

Thus, the marriage of democratic socialism and the appropriate technology movement that is represented by the Citizen's Party, and hinted at by Kennedy and Brown, reflects a radical faith in decentralization and individual rationality.

In the center liberalism and pragmatism—a mixture of left and right—reign. President Carter's critics argued that his mixture appeared as multiple and contradictory decisions, but there is no denying the pragmatic intent of his energy policies.

With the election of Ronald Reagan, however, an administration was brought to power that based its public ideology on the reaffirmation of U.S. power and plenitude. Reaching into the past for its driving images, the Reagan team reclaimed an energy ideology of bounty and exploitation from Old West days. In asserting this vision, they struck strong chords on a theme of considerable popularity.

Whether or not Americans choose to follow Ronald Reagan's ride back into the mythology of boundless energy supplies, the fact remains that important policy decisions must be made. These decisions will greatly shape the energy realities of the twenty-first century. Whichever scenario they permit or choose will affect the daily lives of all Americans into the far-distant future. The next section considers these behavioral implications.

WHAT WOULD IT BE LIKE
TO LIVE IN AN ENERGY-SHORT CITY?

Economically speaking, energy shortfall means that energy prices will rise steeply, such that energy use will of necessity be greatly curtailed. From the perspective of the individual consumer of energy, such a condition will be experienced as an absence of energy sources formerly available. While the economic conception of shortfall is an abstraction, the psychological impact of shortfall is a painful reality.

In the event of energy shortfall individuals will seek to conserve energy as fully as possible while expanding a variety of energy sources that will become available to them. A variety of forms of energy coproduction can be expected to develop, in which consumers become active producers of their own energy supply through direct generation or more effective utilization of solar energy sources. The development of photoelectric cells, for example, is expected to provide individual homeowners with the opportunity to generate electricity on their own rooftops. This technology is expected to be affordable by the end of the 1980s (see Burke, 1981).

For those with the resources to enter the realm of energy coproduction, either individually or in concert with neighbors, living in an energy-short city will be a time of manageable challenge. Some leisure time will be spent in gardening and caring for solar installations. Investments in weatherizing their homes and purchasing energy-efficient appliances and automobiles will quickly be returned, as the prices of nonrenewable energy sources rise rapidly. Food, other necessities, and modest luxuries will remain within their reach, as the well-off adapt to the new era of diminishing hard energy.

Energy shortfall will be more cruel to those of low and modest income than it will be to their wealthier neighbors. All but the simplist weatherizing would require financial resources to be diverted from needed food, shelter, clothing, and medical expenditures. While individual metering of apartments would encourage direct conservation for nearly all Americans, there is no question that the burdens of shortfall would rest heavily on the poor.

As personal transportation costs increase while mass transit becomes increasingly paralyzed, even access to potential employment would become problematic for many. Energy shortfall would be a profound force for the creation of two Americas: one well-off and resource-full, the other poor, cold, and increasingly destitute.

As a nostrum for these ills, coproduction may be seen to offer too little too late. Surely one should not imagine that self-help among the poor will ever replace the many millions of jobs that have been lost to our increasingly hard-pressed national economy. Stokes strikes a balance between the challenges and the promise:

> The transition to an era of greater dependence on renewable energy resources will take time. Consumers experience the vulnerability of dependence on nonrenewable energy resources only indirectly, through price rises and periodic shortages. But the initial capital costs and maintenance problems of solar technologies affect people directly and may dull their enthusiasm, despite the long-term advantages of solar power. Moreover, some solar applications are novel, and people are suspicious of new technologies until they have mastered them. The important thing for consumers is that solar technology is something they can master [1981: 49].

The present author is convinced that we live in a post-affluent era (Gappert, 1979), and that new methods for personal and community survival will be developed of necessity as means of coping with emerging social and economic realities. If we are indeed entering an era of permanent unemployment at or beyond 10%, widespread underemployment, and continuing stagflation, we will need all the food, jobs, and energy savings that can be generated by cooperative neighborhood gardens and solar collectors. Coproduction, while no substitute for economic justice and full employment, may be nonetheless a key in the survival of many Americans in a difficult, and possible, urban future (Van Til, 1982; Van Til et al., 1982).

The energy-short city of the twenty-first century has the potential of being one of the most uncomfortable and violent places human beings have ever sought to inhabit. But it also has the possibility of being a place in which new images of stewardship and cooperation can flourish, a community in which people can not only survive, but rather prosper and grow. The good city need not elude us simply because energy patterns change. But the good city will need to be built, heated, and fueled with foresight long and sound. It is a challenge all urbanists and urbanites would do well to heed, and soon.

REFERENCES

AMARA, R. (1981) "The future field: searching for definitions and boundaries." The Futurist 14 (February): 25-29.

BERRY, B.J.L. (1980) "Urbanization and counterurbanization in the United States." Annals of the American Academy of Political and Social Science 541 (September): 13-20.

——— (1975) "The geography of the United States in the year 2000," pp. 106-136 in J. Friedmann and W. Alonso (eds.) Regional Policy: Readings in Theory and Applications. Cambridge: MIT Press.

BURKE, J. R. (1981) "Photovotaics: down to earth at last." Solar Energy Research Institute Journal (Spring): 4-13.

BURTON, D. J. (1979) "Energy and urban form," pp. 201-231 in G. A. Tobin (ed.) The Changing Structure of the City: What Happened to the Urban Crisis? Beverly Hills, CA: Sage.

COMMONER, B. (1979) The Politics of Energy. New York: Random House.

CORNISH, E. (1979) "The Great Depression of the 1980s: could it really happen?" The Futurist 13 (October): 353-380.

ETZIONI, A. (1968) The Active Society. New York: Free Press.

GAPPERT, G. (1979) Post-Affluent America: The Social Economy of the Future. New York: Franklin Watts.

HARMAN, W. W. (1979) An Incomplete Guide to the Future. New York: Norton.

KRENZ, J. (1980) Energy: From Opulence to Sufficiency. New York: Praeger.

LEVEN, C. L. [ed.] The Mature Metropolis. Lexington, MA: D. C. Heath.

LIPPITT, R. (1982) "The changing leader-follower relationship of the eighties." J. of Applied Behavioral Sci. 18, 3: 395-404.

LOVINS, A. (1978) Soft Energy Paths: Toward a Durable Peace. Cambridge, MA: Ballinger.

SJOBERG, G. (1960) The Preindustrial City: Past and Present. New York: Free Press.

Solar Energy Research Institute (1979) New Prosperity: Building a Sustainable Energy Future. Hanover, MA: Brick House.

SOLOMON, L. (1980) Energy Shock: After the Oil Runs Out. Toronto: Doubleday Canada.

VAN TIL, J. (1982) Living with Energy Shortfall: A Future for American Towns and Cities. Boulder, CO: Westview.

——— (1979) "Spatial form and structure in a possible future: some implications of energy shortfall for urban planning." Journal of the American Planning Association 45 (July): 318-329.

———MARGOLIN, J., and R. CULLETON (1982) Toward Coproduction in the Human Services: Camden College Rises to the Challenge of Retrenchment. Camden, NJ: Forum for Policy Research and Public Service.

WARNER, S. B. (1962) Streetcar Suburbs: The Process of Growth in Boston 1870-1970. Cambridge: Harvard University and MIT Press.

9

Middle Managers Go on Strike:
A Report from 2001

LARRY HIRSCHHORN

JANUARY 2, 2001

Easton on Strike

The Easton Association of Middle Management (EAMM) of the Easton Communications Corporation announced that its members would go on strike at 12:01 tomorrow morning. Seymour Verden, technical analyst of the marketing staff and association president, told reporters this evening that the recently proposed merger between Easton and Western Bell would "significantly disrupt middle-management jobs and careers, and undoubtedly lead to management layoffs as well." The EAMM, said Verden, learned of the merger talks last month and submitted a memorandum of questions to the president's planning group about the merger's impact. In particular, EAMM sought clarification about likely divestments that might follow upon the merger, the implications for staffing patterns in the office and laboratories, and the impact on promotional structures within the corporation. "The president's office responded with a superficial and inadequate white paper that simply assured members of EAMM that their employment and career needs would be fully considered in any reorganization plan. These assurances, to say the least, were inadequate. Attempts to obtain meetings with the president and board were rebuffed," said Verden, "and we had no choice but to poll our members on the possibility of a strike. Ninety percent of our members voted to support the walkout." When asked how long he expected the strike to last, Verden smiled, "It depends how much we can hurt them. We have little strength out in the rural branches and I don't expect support there. But Headquarters Operations is central to the sales, marketing, financial, and planning operations of the company. We can shut it down. The stockholders, particularly the representatives of Asian governments, will yell 'ouch.' "

The president of Easton, Henry Barton, was unavailable for comment. Jack Dooley, public relations representative for Easton, released the following statement to the press. "The strike declared by EAMM will not disrupt Easton operations. All employees are expected to come to work tomorrow. We expect that all our plants will be in full production."

It is difficult to estimate the impact of the walkout on Easton operations. Prior middle-management strikes in high technology have not been successful. But EAMM is known to be a militant professional association and may have developed new plans and tactics.

JANUARY 7

Marketing Division Walks Out

Offices of the marketing division of Easton Central were empty save for the slave station secretaries at the voice-processing centers. Upper management was shocked to discover that analysts have joined the EAMM strike. "I thought for sure they were loyal to the company plans and policies," said one manager who asked that his name not be used. "It goes to show how out of touch Barton was with this pet division. He imagined he was grooming future leadership from marketing and they joined EAMM. Only a few were EAMM dues-paying members and I'm sure that none went to EAMM meetings. It is a real shock."

Seymour Verden, president of EAMM, met with strike captains at the city rest house and haven and noted that "marketing staff had been put through the mill of uncertainty over the past year and a half. You can't believe the rumor mill that was operating in the office for the past eighteen months. One day people were sure the board was pushing divestment strategy, the next day an acquisitions strategy, and the next day a merger strategy. It became impossible to do any meaningful business or personal planning. One guy was developing a voice-actuation product for Dupont's very noisy underground farm mills. He spent four months getting to know the industrial process and had numerous meetings with the plant managers. One day he walks in and the Dupont plant manager tells him that he hears Easton is about to divest of its glass fiber and transmission plant. He didn't want to pursue negotiations further until the situation was clarified. The rumors were spreading throughout industry. The crazy thing was that Easton staff could not get good and hard information from upper management. Upper management screwed itself by pleading the need for secrecy. When are these guys going to learn to involve the staff at least minimally in the future planning of the company?"

Verden went on to describe the situation of others who were unable to make any decisions because of the uncertainty. "One guy kept asking me, 'I want to place my mother-in-law in the university nursing home near my apartment, but will I be living there in a year's time? I can't put her in and take her out. What the hell am I supposed to do?'"

An informant who has the ear of upper management could not clarify the potential impact of the marketing staff walkout. "It could really kill negotiations that are critical to next year's sales. We'll be able to process orders and meet maintenance calls, but new business is always the bulk of our products and profits. This could upset a lot of customers."

JANUARY 15

News Analysis: Behind the Strike at Easton

The strike at Easton is one more sign of the growing disorganization of corporate life in the Fortune 200. The Easton Association of Middle Managers (EAMM) first emerged some ten years ago when a woman's pressure group at Easton opened its doors to other dissatisfied management employees. Initially, the women's group (Women Managers of Easton—WME) was formed to protect and enhance the careers and promotional opportunities of women at the company. In the eighties WME played an unobtrusive role, providing counsel to its members on the tactics of survival at Easton. It pressured EEO officers and employee relations people to monitor the impact of promotion decisions on women's status in the company.

WME was greeted with hostility by many men, but the rapid growth of the communications industry itself, particularly the voice component, gave everyone the promotions they wanted. Career lines were plentiful. Then things began to change in 1990. First, in the eighties many women emerged in management through the office automation revolution, often beginning as secretaries, then moving up to office manager positions, and finally into line management positions. Ironically, women MBAs who thought they had it made were not given the chance to prove their mettle in tough operating positions.

They were shunted aside to the soft areas like human resource development. Many of them viewed the results ambivalently, because the nonoperating jobs did give them time to be mothers. The masculine culture at Easton is unforgiving and once a woman showed that she liked those "nurturing" human relations jobs she could never get back on the core operations track.

The completion of the first wave of automation by 1992 slowed down the upward mobility of the tough women managers and as a consequence the proportion of women in middle- and top-management jobs began to fall. In addition, competition for entry-level and middle-level communications industry management jobs intensified in the 1990s. Many men from other declining sectors, born in the much-described baby-boom of a half-century ago were looking for second careers in the communications sector. Their own career lines had been blocked, but with declining numbers of new entrants into the labor force, some took a chance and bid for entry-level jobs in the communications sector.

Shelly Fuchs, vice-president of EAMM, reflected on this period. "It was strange to see these men in their mid- to late-forties taking junior positions. I never thought it possible, I didn't see how they could afford the cut in wages. Apparently, many had wives who were making good money, most had only one kid, and the singles had few expenses. All were sick and tired of being stuck at the bottom of a career ladder in a company that was going nowhere. The university intension courses that developed in the late eighties seemed to have an affect. For a while I thought they were simply a mishmash of life-planning, t-groups, and positive-thinking methods, but they turned some heads around. Men took chances they never dreamed of taking before."

"I remember seeing a cable documentary from the 'History as Present' Center. It reminded me of the crazy sexual atmosphere that emerged at the time. Let me call up the tape. I dug up its number the other day." (A video clip comes up on the desk screen. An attractive woman in her thirties named Marla is talking. A vintage Radio Shack processor is clacking away in the background.)

It's a bit crazy, the rules of sexual etiquette are all awry. If I invite my older male subordinate to lunch to talk over business, who pays, who orders, who sits where, do we shake hands? You know, sexuality and authority are very tied together. Remember the notion that power is sexy? I have the distinct feeling that some of the divorced men are asking me to seduce them. But if that is the case who is really doing the seducing? As to the competition between younger men bosses and older men subordinates...the competition is intense, and sometimes I feel that I'm the sub rosa object of that competition. There is a crazy sexual game going on underneath the smooth surface of corporate life. Only pressure from the top for performance and the people's own instincts of craftsmanship keep the ship going.

Shelly Fuchs concluded by saying, "Warnings at the time that Easton had to develop a human relations program were ignored. Easton upper management was only too happy to pay entry-level wages to experienced managers in other sectors. I am sure that this crazy atmosphere had something to do with the emergence of EAMM. People felt a little nutty and they didn't know why."

In 1990 WME made a critical decision to open its doors to all managers, feeling that only in this way could it retain its political strength in the face of the new cohort of management employees that was entering the company. After a year of exploratory meetings with many male managers from different divisions the WME changed its name to the current EAMM.

The strike at Easton is a watershed in the history of company-employee relations. Other strikes in earlier years, such as the famous one at Exberg when corporate monitors walked off the job, quickly collapsed. Since then, however, there has been a significant growth in informal middle-management pressure groups throughout industry. Whether these pressure groups will ultimately become the new unions of our post-industrial economy has yet to be seen.

JANUARY 20

The Scene from the Office

The middle-management strike at Easton is in its third week. Stockholder pressure on the board is mounting to simply fire all of the employees participating in the walkout. Upper management is reluctant to take such a drastic step. One informed insider noted that "upper management is in a real pickle. If only our monitoring staff had walked out we could do what the president did at Exberg—fire the bastards. But we are talking about our marketing and sales people at headquarters. They constitute half the value of our 'good will' with our customers. I just can't see firing the lot and risking the loss of a decade of good customer relationships. Besides, at this point it is hard to know who to fire. Some managers have taken advantage of our flexiyear policy and have taken study-leaves. So they are striking with pay for all that we know. Our employee rights officer tells us that we shouldn't risk suspending flexiyear pay. The corporate courts might not uphold us. There must have been some secret planning here, because a larger number of requests came in some four weeks before the strike. Frankly, the duplicity disgusts me.''

When asked if the merger with Western Bell might not fall through, he noted, ''Bell is patient. They've been going through their own middle-management crazies over the last decade. Things have never been the same there since the breakup of the company some eighteen years ago. They also know that our value to them falls if we can't settle this peacefully. We have just got to keep the stockholders off our backs. The stock price has dropped some 10% since the beginning of the strike. Some of the Indian banks are getting nervous. Of course Bell might pull out at some point.''

Thus far there has been little public acrimony between the officers of EAMM and the presidential staff. Clearly, upper management is doing all it can to keep operations going. One secretary describes the scene. ''I'd say that about 80% of middle-management types are not showing up. Some vice-pres types from the eighteenth floor have come down to run the various departments. They don't know anybody's name, can't locate things on the terminals. The secretaries are laughing and the high school terminees, particularly the men at the voice stations, are doing some dirty tricks. But the VPs shouldn't worry about us small fry. We can manage the offices. It's the decisions with the new customers, suppliers, banks, and credit companies that must be made. That's the real problem. Seems like the word is to postpone decisions where possible. It is becoming a real circus. I don't know how well the banks will stand for this.''

JANUARY 27

Two Strike Votes at Easton Plants

The United Electrical Workers of American voted today to support the strike of middle managers at Easton, but the Organization of Atomic and Chemical Workers voted not to support the strike. This means that the production of fiberglass at three southern Seattle plants will continue. But production of the new tactile message system at the Vancouver plant will stop. This is interpreted as a victory for management since the fiber plant contributes most significantly to the annual profits of Easton. The tactile message plant is Easton's newest venture and is not expected to turn a profit till 2002. But certainly, Easton top management is banking heavily on the productivity of the Vancouver plant.

Jack Tooney of the United Electrical Workers explained the reason for the difference between the two votes. "Basically, the people in the tactile plant identify with middle management. It's a high-class experimental factory, basically like the *labactories* at University Station here in the city, and the Easton company has insisted from the start that all workers have a BA or more, preferrably with a background in some technical or scientific field. The work is demanding. The tactile-visual stuff, you know, is used a lot for telemedicine work and touch sensitivity has to be translated with great fidelity between any two locations. The doc can't be feeling a head when in fact he is feeling a toe. The computers do a lot of the production legwork, but the adjustments to the production process are continuous, particularly since we make the stuff in small batches. Word around the plant was that if you wanted to make it into management without a business background you should start at the labactory and really get to know the technical end of what should be a growing part of the business.

"But the guys at the chemical plant, they're fed up with middle management and all their shennanigans. Basically, plant managers and their support staff, a lot of whom were looking for promotions into central headquarters, would constantly mess around with the shift work arrangements. For the last five years management could not develop a smooth work shift system. Each time production specifications changed, and they changed as the number of international orders increased, they had the guys doing a log of code changing and endless retesting of the circuitry in the control equipment. Then, as one foreman tells it, the simulator would start to crank out crazy numbers on the feed-forward program. The guys would complain but the computer woman insisted that the gas-screen was simply malfunctioning in some way. Apparently she was hot on the old CRTs and collected them for a hobby. Anyway, management simply couldn't predict the staffing patterns required and soon the workers found themselves on a three-day-long night shift once every two weeks. It was intolerable. The shop stewards first brought up the problems through the scientific circles and study meetings, but the management reps dodged the questions. The stewards then brought it up at the grievance committee but middle management kept pointing the finger at top management saying that headquarters simply refused to add capacity. But the reps didn't believe them. They seemed to be just covering up for their own incompetence and when pressed, they kept going back to the national contract, which gives management total prerogative over shift work allocations. The workers are well-paid but were getting disgusted. The national would not have supported a strike over shift work, so refusing to support a middle-mangement strike was the next best thing to do.

"As to the unions in the other plants, the old video-recorder plant, the picture phone plant, the copper cable plant, and god knows how many others, I don't think they even know a strike is going on. After the big decentralization of ten years ago the plants run pretty autonomously. The ordering process is practically automatic based on depreciation formulas Easton worked out with users over fifteen years ago. Maintenance is subcontracted to the high school consortiums around the country. With labor content so low, synthetic materials content so high, and component reliability 98% the sales process is basically computer-managed. The plants are just evaluated by headquarter monitors and have to meet annual financial objectives. Anyway, a lot of them are out in the new rurals of Detroit, Philly, and the like. And those are the boondocks!"

FEBRUARY 1

News Analysis: The Strikers: An Inside View

The apartment of the strike captain, John, is modestly decorated. The information screen, a deep red when in the lock position, is placed unobstrusively against the partitioning fabric. The lighting pattern follows the familiar textural rhythms of the Cy-three computing system. Dark grays play counterpoint to bright yellows in a dance that soothes. Pictures of bag-children in Eastern St. Louis, Camden, and Miami hang on the wall to the side of the information screen. Their abandoned look contrasts strikingly with the organized appearance of John's living system. John had worked with the bag-children once through the Christian salvation movement. His niece had been dumped by her parents when she was ten and she became a bag child. John wondered why she never was drawn into the marauders. Her chronic rage colored all her movements and expressions. A group of strikers is talking. A reporter listens.

JOHN: I looked at some pictures of the great strikes of the nineteenth and twentieth centuries. I know we're not fighting for bread. But you've got to understand the stress, the crazy-making atmosphere in so highly politicized a company like Easton. You lose perspective, you can't do a good job anymore.

MARY: We're all survivors. Tomorrow we could go out and sell credit services to the Russian immigrants at the jitney shops. We are not going to become victims of the system. Corporate leadership doesn't understand how free-floating we've become.

BILL: I became free-floating in my mind some time ago. I was on the fast track, all hep to work in the tactile-marketing division. I don't know what happened but somehow the whole enterprise stopped making sense. I think I felt betrayed when the company shifted its commitments from the laser technology. They were taking the easy way out, the quick payoff, instead of developing a product that could have led to great things.

(The reporter tells the group that they look and feel resigned. "Is this the elan that produces a strike victory?" he asks.)

JOHN: Resigned? Well, no, well yes. You're right in a way. The strike is a funny thing. We're doing well financially of course, and as many as a third of us have kept our family life together. Again, those pictures of the nineteenth- and twentieth-century strikers are suggestive. Those people wanted a fair share and to be treated with dignity. No one is going to make me feel dignified or not dignified. Today corporate leadership cannot shape my feelings about myself the way in which nineteenth-century leaders shaped the immigrant's picture of himself. God knows our upper class today can't defend its commitments or behaviors to itself. Sometimes I think that they too have resigned, in the existential sense, if you know what I mean. Often they are just playing a game, they've stopped living in the real world in a way. They live with simulators most of the time, they change "parameters," as they call them, at the push of a button. Somewhere inside them they have stopped believing in the existence of a real world. In a way, our strike is to remind them and us about that real world. It's a funny reason to go on strike.

JUDY: It's funny. We want to "participate" in the company, as Verden says (everyone smiles), but in another way we want to get in so that we can get out on our own terms when necessary. Basically, we need to get in so that we can stop them from stopping us from doing our own thing. Is that a reason to strike?

FEBRUARY 8

Easton Shifts Managers

The Easton Corporation in a surprise move has shifted two hundred managers from its rural divisions into headquarters to take over critical management marketing and financial functions. Easton has contracted with the Simulac Corporation, a major consulting company, to organize the entire effort. A top company official told reporters that "by placing some of our best managers in key positions held by the strikers we are able to insure the continuity of our operations. The banks in particular should rest assured that we can meet all our obligations in time." He then handed out a press release from Henry Barton, president of the corporation.

"Simulac, Incorporated has agreed to oversee our headquarters operation for the remainder of the strike. I am warning EAMM members that attempted sabotage of company operations will be dealt with severely. The corporate courts have been kept abreast of all events and their administrative panels are ready to accept a quick increase in strike-related cases. Simulac's leadership has assured me that its staff will be monitoring computer usage. Any attempt to break into the decision support system will be quickly detected."

Seymour Verden, president of the EAMM, appeared shocked. "Simulac is a notorious consulting firm that has played a very destructive role among the Fortune 200 in the past. They were behind Winston's divestments of its agridrug facility on the grounds that projected political trends on a global basis dictated a different pattern of long-term investment. This despite the fact that the facility was very productive and top management had made ten-year contract commitments to its personnel and workers. I needn't tell you the rest of the gory story. They kept the divestment plan a secret, milked it for profits without maintaining or enhancing the information structures, and then dumped it into the open market where a banking subsidiary actually bought it for a song. And then, to top it all off, after the Frisco earthquake hit Simulac turned all its political predictions upside down. I am shocked and angered that Easton has stooped so low. They have picked true postindustrial pinkertons."

A union insider with ties to lower-level employees at Simulac noted that "these consultants are a weird bunch. They remind me of the computer junkies I meet at those game parties, awkward, cerebral, afraid to ask you to dance. They're all loners, even the married ones, and they move with the job. I think in some funny way they've been hypnotized by their own data and decision systems. The feedback loop has turned in on itself and they act as if they've been programmed by the computer they've programmed. It's a little like the schiz who talks to himself on the skyways, except that Simulac people get paid big money for their monologues. But there is no doubt that they are very expert at information systems protection. They have devised monitoring systems to track usage and are able to create bait programs that pull unauthorized users into the warning alarm loop."

An EAMM sympathizer noted, "The trick will be if EAMM types can get into the credit program. We tried it at a company reorganized by Simulac some three years ago. We lost."

FEBRUARY 14

Bank Crisis Averted

A near bank panic was averted today when the Federal Reserve Bank of Vancouver announced that it will cover all the unexpected short-run debits and obligations of the large First National Bank of the United States located here in Vancouver. In an off-the-record comment, one informed source noted that "the whole Easton affair is clearly behind this." The story has yet to fully emerge, but apparently First national officers would not disentangle the flow of funds between creditors and debtors in its regional information system for the past week. At first, reserve officials suspected bank incompetence, but a careful analysis of past printouts and a simulation of annual flows based on last year's data indicates that critical coefficients had been tampered with. "It seems as if," said one bank official, "that positive and negative coefficients had been reversed in a few concentrated components of the regional model. It was too unrandom to have been caused by circuitry problems, particularly when you note that the greatest confusion in flow of funds calculations is concentrated in the 'Easton-to-other' sector. If the Feds hadn't stood behind us we might have decided to go to Easton and demanded immediate liquidation of property to back up prior outstanding loans. Otherwise we might have gone under ourselves."

A spokesman for Harvey Bender, president of Easton, said that "Easton Corporation and its board find the obvious and attempted sabotage of its financial interface with the Federal Reserve system to be reprehensible. EAMM should know that our resolve has strengthened, that we will risk our merger with Bell rather than hire back computer criminals that now populate EAMM ranks." Seymour Verden, president of EAMM, noted that "its association could take no responsibility for the actions of a few individuals who may have sympathized with EAMM members and sought to put pressure on Easton, its officers, and its stockholders. We again ask Easton to offer a framework for negotiations within which the strike could be settled."

Simulac had no comment on the bank affair. An informant noted that "this is an embarrasment for Simulac. Their reputation depends on ironclad protection on information systems. Obviously they didn't think to look into the internal structure of the banking model itself. Nor is it clear that they could have gotten permission. No scheme is 100%"

FEBRUARY 29

Strike at Easton is Dying

The middle-management strike at Easton Corporation is slowly dying as an increasing number of managers are returning to work under a unilateral amnesty policy offered by Easton's top management. Most will not lose positions or career ratings in the company, but none will get back pay. "It's a fair deal," said one returning manager. "The near bank collapse really scared me. I guess I wasn't ready for that level of disruption. After all, my credit line is in that bank and I was just notified that I qualified to open a life-credit account. You can't risk bringing down the system you know. After all, we are also part of the system. Can we strike against ourselves?"

Seymour Verden, president of EAMM, and loyalists around him refuse to call an end to the strike. "It's true that our support is eroding, maybe dying, as some people say. But it is key that we at EAMM make a symbolic last stand. I'll lose my job, but I'm a survivor and the principles of our strike, the right to plan one's personal life, the right to company information, the right to a modicum of certainty is one's life, these have to stick in people's minds."

"So the company thinks it has won. But did it really? The managers go back but the climate is the same, the level of dissociation is the same. People are not going to give their best technical and financial ideas to top management. You saw how ready a large number were to rip-off Easton by striking and taking flexiyear pay at the same time. The rip-off mentality is going to hurt everyone in the long run. If this climate persists, we may all become psychological clones of Simulac consultants. Maybe we should do what they do, live inside the information systems we ourselves create and lose touch with human behavior. Will this make the North American economy great? Only if you start making up your own GNP data and start changing the statistics books, the government reports. But why stop there? Why not change the encyclopedias, the history texts? Is this where we are headed?"

INTERPRETATIONS

Place: Philadelphia, General Development Corporation, Headquarters. Game Room seventeenth floor.

Activity: Management Development Program.

HENRY *(management development specialist)*: Let's turn to the Hirschhorn article. I gave it to you to give you a flavor for one kind of scenario writing style that emerged in the eighties. What did you get from it?

BOB *(VP careers planning)*: It was fun but funny. Story writing about the future can be a seductive undertaking. I see how Hirschhorn got carried away by the logic of his own story and lost contact with the underlying trends that motivated the scenario to begin with.

JACK *(line worker, rotating through management committee)*: What do you mean? I liked the story, I thought it described a lot of undercurrents here in our own company. The description of the chemical plant was eerily close to some of the problems we're having in Wilmington right now—particularly the shift issue.

BOB: that is what I mean—undercurrents yes—but actual facts and events—no. For instance, the middle-management strike is a cute idea but looking at it from today's perspective it seems a bit far-fetched. Hirschhorn missed the resilience of corporate culture, the adaptability of management in general. Its true that the downsizing of middle management throughout the corporate sector has had great repurcussions on careers over the past twenty years, but at the same time it offered great opportunities to many managers. They have more flexible careers and the freedom to job-hop. The strike idea limited Hirschhorn's imagination. He could see development only in the context of nineteenth-century notions of conflict and dialectics. That is not the kind of world we are living in today.

WILL *(plant manager, biochemicals division)*: I'm not sure I understand Bob, but it seems to me that strike or no strike he hit at some central themes. The issue of "participation" stood out for me most—you know, Verden at the end arguing for the employees' right to participate in the plans that affect their futures. I feel it today, it's sometimes so difficult to deal with the turbulence. I can't tell you how many times I have withheld information from my direct reports, information that could affect the sale or shutdown of a key plant. It kills me.

JOHN *(university liaison)*: Actually, I thought that the theme of participation was more subtly treated. Remember the scene at the strikers house? The union members are ambivalent about participation and the woman Judy says, "We want to get in so that we can get out on our terms." That struck a chord for me—I often feel that the call for participation today lacks a positive thrust. People want to check the decisions of leaders so that they can maximize their own freedom to make or break commitments as they wish. At times I feel good about this, I see it as a strong anti-totalitarian thrust in our culture. But at other times, it makes me worry. Where is the glue that holds us together? It seems as if we have to wait for unexpected erruptions, May 1968 in Paris, April 1988 in Los Angeles, to really feel our collective experiences and potentials.

BOB: Paris, May '68! Los Angeles, 1988!—you university liaisons are still dreamers. When will you give up the wish for social eruptions? I took the labor-management history course last year, I can appreciate your nostalgia. But that is all it is.

HENRY: Bob, let me expand on your point—it touches on my concept of scenario writing. Its true that Hirschhorn may have been stuck in some old concepts, but the really important issues are how he uses the story to elicit a complex theme. Stories, unlike scientific papers, are based on a flexible narrative structure. They can suggest contradictions and tensions without appearing illogical or incomplete. Instead, the contradictions invite the reader to fill in the story, to respond both affectively and cognitively to manifest and latent themes. The strike is the story line, it allowed Hirschhorn to get on with the business of constructing characters, but the real message lies beneath the events. I thought that John hit on one key theme here: the ambiguity of participation. The strike metaphor allowed Hirschhorn to develop this theme without lecturing to us. I believe that this interactive role of the scenario, the way it invites participation by the reader, is its most essential feature. This is why it has proven so useful for management decision making. It complements the abstractions of data-based decision making by giving play to judgment and the sense of ambiguity.

BOB: I see your point, though I think in applying these methods here in GD we can't get too science-fiction on the one side, or backward-looking on the other. Even if the story line is handmaiden to the underlying theme, too artificial a story line will distract the reader. I found the allusions to "bag children," "corporate courts," and the like confusing.

JACK: I liked those. God knows there is an anti-child crusade in the culture right now.

(A secretary interrupts and comes rushing into the game room)

JEFF *(secretary)*: Bob—the president is on the phone. He's panicking. The situation in the finance department is blowing up. We can't afford disruption now, he says. He needs you to bullet-train to New York immediately.

(Bob exits with a hurried goodbye, leaving his papers on the desk.)

HENRY *(seems oblivious to Bob's exit)*: Let's explore this theme further. Look at Hirschhorn's 1986 family scenario....

10

New Technologies and
Their Urban Impact

JOSEPH F. COATES

◻ LOOKING BACK 50 YEARS to the Great Depression, one finds a great deal of visual, structural, and functional continuity in the cityscape. Looking back 75 years to the first decade of the century, there is much less continuity, since many key inventions of the previous decades—automobiles, electric lights, steel beam construction, telephones and subway and trolly lines—came into full flower between 1915 and 1940. Casting our minds forward 50 to 75 years, we can expect similar patterns of change. The size of sunk investment, natural turnover times in urban structures (about 40 years), and now-embryonic technological developments will present a diminishing but real pattern of continuity with change.

This chapter scopes some of the most likely changes in the next 50 to 75 years as they are driven by new physical technologies. There will be no typical city. There will be new cities in new locations. Some present cities will so prosper and expand that they will seem to change completely. Many older cities are on the wane or are moving to some phase of maturation and stability. One strategy for getting at this mixed future is to look at seven different ways in which technology is likely to change cities. These seven routes to anticipation of effects have solid historical bases, and hence strong appeal as ways of understanding the future. These approaches consider:

(1) *Structural effects of technology on the economy.* Extraction, transportation, and manufacturing technologies are the backbone of the industrial society. Pittsburgh, Youngstown, Detroit, and Akron, for example, were largely determined in location,

organization, and function as symbionts of the steel and automobile industries. As these industries wane, or manufacture moves elsewhere, those cities are likely to decline. Similarly, the growth of agribusiness and the shift from family to industrial farms wiped out many of the smaller cities and towns. Christaller's central place theory explains the decreasing number of communities and the changes in the hierarchy of city size by the expanding geographic range of social and economic functions provided by modern communities.

(2)   *New technologies will affect the organization and structure of the city.* Among the dozen obvious examples in the past 75 years are the electric light, which gave us an active nocturnal city; the elevator, which made high rise buildings practical; and the automobile, which has given us so many things from mobility to sprawl to pollution to accidents.

(3)   *Technology as a response to problems.* Technology often develops in response to problems. River and coastal flooding have had substantial effects on the location and design of cities. Certainly, new rounds of knowledge will continue to promote developments along these lines.

(4)   *Technology has effects through what it displaces.* Technology is almost always introduced as a substitute for some thing or some function. The introduction is driven by the desire for efficiency, effectiveness, or some other micro-techno-economic advantage. Side effects often accompany the supplanting of an older technology. The automobile, in displacing the horse, wiped out the urban horse manure problem. A direct analogy in the future may be better crime alarm and safety systems. That would reduce the need for dogs and wipe out the dog droppings nuisance.

(5)   *Technology interacting with other trends.* This, of course, is the most speculative and uncertain of the routes to understanding technological effects on the cityscape. Urbanization and industrialization have led to more densely populated cities, with the structural consequences that transportation systems and food and water distribution systems reflect that density and become overelaborated.

(6)   *As technology becomes institutionalized, it will evoke other technologies.* For example, movies as entertainment led to the movie house. Technological developments making practical large-scale food preparation, coupled with the increasing percentages of people working away from home, led to fast-food emporiums and fast-food chains. The new institution further promotes technology for a high-quality, reliable, uniform food delivery system.

(7)   *Special Cases.* Most difficult to anticipate, but often having dramatic effects, are unique or community-specific develop-

ments. Some boom towns in the Rocky Mountains boomed and busted several times in a century. Los Alamos was selected as a remote area for highly secret military R & D.

The rest of this chapter pursues these seven routes to the future of the city. The emphasis is on developments in the United States likely to result in macro change. Only a few of the more conjectural developments will be mentioned that may give the local texture and diversity so visible to the eye. The discussion is limited to physical technologies, but we call the reader's attention to the fact that every physical technology must be part of a social system, which itself will evolve to frame an institutional context and system for the physical developments. The organizational context will itself call forth new physical developments and social inventions.

Now, back to the seven themes and forward 50 to 75 years.

THREE BIG STRUCTURAL CHANGES

Three major structural changes in the national economy that will have derivative effects on most cities are: (1) the movement to a postindustrial society, (2) the internationalization of the American industrial sector, and (3) the irreversible increase in the cost of energy.

The movement to the postindustrial society, a term coined by Daniel Bell, marks the transition from a society in which industrial production, physical goods, and material property and their ownership and management were the dominant themes. The new postindustrial society, in contrast, finds knowledge-based and knowledge-driven industries dominant. This is reflected most sharply in that the bulk of the labor force is now predominantly in the information business and only 22% is directly engaged in industrial production, that is, manufacturing, and only 3% in agricultural production. Much of the social and physical infrastructure of American society is premised on an industrial society. Everything premised on industry, from the patent system to inheritance laws, from manufacturing plant siting to highway design, is on the road to obsolescence, insofar as the postindustrial society calls for or permits a different reorganization of physical space. This transition will effect the city and the reorganization of the built environment because it makes obsolete, unnecessary, and wasteful the high degree of centralization required by large-scale manufacturing.

The postindustrial society also makes obsolete the large-scale highly centralized white-collar work towers of every downtown in America, which grew up as adjuncts to and in mimicry of the factory. The reorganization and dispersal of those white-collar work towers will not come about simply because the original justification no longer exists. Drastic changes in white-collar work will come about for reasons elaborated below—the technologies of telecommunications and computation now make it practical and desirable to disperse that information work force.

The second big structural force is the internationalization of industrial production and sales. The U.S. industrial giants whose good fortunes have been based upon manufacturing will continue to manufacture. But much of the productition will be in overseas plants. Research, development, design, marketing, and management, at least in the early phases, will be controlled in the states. Some second-generation effects of this outmigration of industrial manufacture on a worldwide scale are already visible. First the United States, then Japan, now Korea or Poland, dominate shipbuilding. The world car is not merely an advertising slogan. It is the label for a new production system in which parts produced anywhere on the globe will be so fully integrated that a car can be assembled anywhere from those parts.

The third structural drive is the sudden sharp rise in the price of energy. By all reasonable conjectures, it will not reverse itself to any significant extent. Energy-extravagant industrial facilities, domestic facilities, and urban structures premised upon cheap oil are now wasteful designs that must change. Distinct but overlapping waves of compensation will occur. First will be retrofit to plug up and compensate for the energy wastage. Second will be the rolling in of new energy-conserving designs. Small units of work and living space, new heating and cooling designs, greater flexibility, and new material will come into play.

These three structural forces add up to different outcomes on a community-by-community basis. Many cities will go the way of Youngstown and the New England cotton mill towns. Industry will dry up or move out, and many cities, for want of foresight, will, like East St. Louis, slide into gross deterioration and become derelict cities—urban ghost towns. Others like Pittsburgh will weather the transition by becoming the regional centers for new economic development in the postindustrial economy. Each city must look to its own future because for many local governments, planning the downside of the urban growth cycle will be a new, frightening demand. The common tendency will be resistance or denial because elected officials are totally unprepared for decline or stability. The center cities, of course, could become far more attractive as space opens up and the white-collar work force is dispersed. But that cannot come about unless there is radical change in planning and in the tax base to promote downside development instead of downside decay.

NEW TECHNOLOGIES
AFFECTING THE CITY

Telematics, that is, the technologies of telecommunications, computers, and information, are the central physical technologies of the postindustrial society. These technologies, by affecting the production, storage, handling, use, and dissemination of information, are central to the knowledge-based economy. Their primary effects will be in the short run on the functions of the city, and in the longer run on the structure and organization of the city. The family of technologies noted in Table 10.1 and shown schematically in Exhibit 10.1 have the capability of dissociating work from the calender, the clock, and any fixed location. Technologies for information generation, analysis, and storage make it possible to do much of the routine white-collar data handling and data processing any place, because inexpensive telecommunications linkages will be able to carry the information to any other central or distributed place. Even more sophisticated functions such as research and analysis can be carried on at sites remote from any central location. The results one might see in the short run would be movement away from centralized office space and to suburban locations and even to homes and specially designed small work sites. More exciting than this centrifugal tendency is the implication that we are on the verge of the truly wired city. My estimate is that by the turn of the millenium, the average household will have an investment in telematics roughly equivalent to its present investment in the automobile. The recreation room is a thing of the past. The home work-study center will be an around-the-clock focus of the household, used by Mom and/or Dad for work, used by the kids for general or special schooling, used by everyone for entertainment and socializing with the outside world. Telematics will put large cohorts of the population in voice-to-voice, face-to-face, or data communication with almost anyone at almost any place at almost any time. How the details of this work out are, of course, subject for major speculations, but the overall trend is undeniable. Much of the concern about isolation and anomie is misplaced. These technologies will be broadly socializing. They will permit more contact with more people for more purposes, and probably stimulate the desire for more and diverse flesh-to-flesh communication. In any case, the structural

Table 10.1

TELEMATICS DEVICES AND SYSTEMS

Telematic Devices Widely Available Before 1960	*Telematic Devices Widely Available by the 1970s*
Telegraph	Touchtone pad (push button
Telephone	keyboard)
Microwave transmission	Call forwarding
Radio (AM/FM)	C.B. radio
Television (B/W	Picturephone
Facsimile	Portapack video recorder
Phonograph	Color television
Tape recording	Portable television
Xerography	Frame grabber
Cable (one-way)	Slow scan
Transponders	Private microwave transmission
Typewriter	Audio cassette
Movies	Low cost xerography
Still Photography (B/W, color)	Two-way cable television
Polaroid	Electric typewriter
Microfilm	Geophysical Satellites
Robots - first generation	Optical scanner
Mainframe computers	Microfiche
	Robots - second generation
	Hand Calculator
	Minicomputer
	Central processing unit memory

Telematic Devices Now & Emerging in the 1980s	*Some Significant Systems & Software*
Microprocessor	Micro home information system
800 numbers	Computer utilities - Illiac IV,
Voice answer back	Arpanet, Plato
Voice activation	Communication satellites -
Low cost video recorders	Comsat, Intelsat, ATS-6
Video discs	Other space Satellites -
Large Screen television	ERTS, weather, agriculture
Laser	CAD (computer assisted design)
Fiber optics	CAM (computer assisted
Video cassettes	manufacturing)
Electronic scratch pad	PBX self contained
Direct satellite broadcasting	telephone exchanges
Word processor	Simulation-modeling
Graphic & color display	Aids to the handicapped
Speech compressor	Electronic switching systems
Packet switching	(ESS)
Robots - third generation	Mobile cellular system

Table 10.1 (Continued)

Telematic Devices Now & Emerging in the 1980s	Some Significant Systems & Software
Microcomputers	Information utilities -
Large scale integrated circuits	The Source, Lockheed, Prestel
New memory systems: solid state	Teletext
laser, bubble, backend	Videotex
processes	Pattern recognition -
Morpheme generator	voice, signature
	Encryption

effect of the office of the future would be to drastically modify the present centralized structures. Corporations now trying to modernize with the current generation of office-of-the-future fixtures may be on the wrong track. The real future may lie in decentralization and work at home.

The facilities that will work for the industrial and business community are the same kinds of facilities that will put all kinds of data and information at the disposal of citizens and thus open up the political and policy processes to more effective intervention. Democracy at the local level will get a great shot in the arm from this technology. For example, with a simple $1200 minicam anyone can prepare their own programs on community issues. As I see it, political good sense will force government to accommodate this new technology in their outreach for participation and in the actions they take.

Telematic is not the only primary technology comparable to the electric light and the elevator. New developments in engineering such as tension structures or kinetic structures may radically alter the basic design of buildings. Most buildings are designed to depend upon mass and rigidity for their stability and their resistance to environmental stress such as hurricanes, storms, and earthquakes. Recent developments now make it possible to build buildings having structural elements in them responding in real time to the stress of wind, storms, and earthquakes, thereby making it possible to build lighter, more airy structures. This opens up possibilities for new uses of enclosed space. This potentially radical shift in construction technology is roughly equivalent to the development of prestressed concrete and the steel beam.

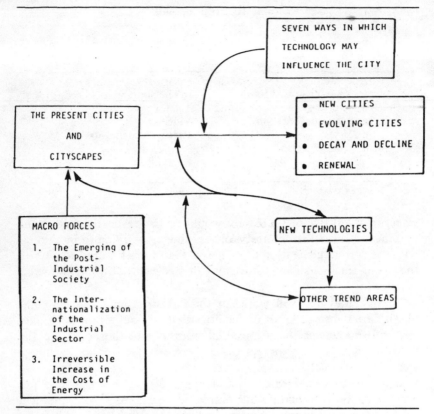

EXHIBIT 10.1   New Technologies and Their Urban Impact:
A Conceptual Map

TECHNOLOGY AS A
RESPONSE TO URBAN PROBLEMS

The cities' underground infrastructure, including water and sewage systems, and the highways and road systems are in disrepair and are obsolete. Some cities will allow the hidden essentials to continue to deteriorate and will follow a short-term strategy of deferred maintenance and piecemeal rehabilitation. Other cities will see an opportunity, particularly in newly redeveloping areas such as the Bronx, New York, to innovate, combining and integrating much of the underground infrastructure for economy and the ease of maintenance. Water supply, sewage, and runoff will continue to be subjects of urban action. It is unlikely that any particular water-related technology will show much in the way of generally visible effects. However, one can anticipate that telematics and electrical networks will make more effective use of underground space. As fiber optics replace copper wire and coaxial cable, telematics will be just a routine part of the utilities of all structures. One can expect plug-in telematics facilities in all structures just as routinely we now have plug-in electricity. A potential problem that may accelerate and promote the use of fiber optics is the continuing uncertainty about the health effects of microwave radiation. Should that become an important public health problem, much of the microwave broadcasting system will be replaced with conventional cable and with fiber optics, which are fully capable of the task. Others argue strongly for reserving the microwave spectrum almost exclusively for satellite and mobile communications and moving all other point-to-point communications to underground systems.

We can be reasonably certain of the general trend in dealing with outdoor air pollution. But indoor air pollution is a worsening problem. Office machinery generates ozone and other by-products. Plastic-covered walls and furnishings exude plasticizers and un-polymerized components such as formaldahyde. As we move toward energy conservation, lower rates of air exchange, and more high-tech interiors, the air pollution problem will become more important. This might lead to quite radical effects in design, air conditioning, and the allocation of indoor space. It is too early to say what paths that will take, but it will be important, troublesome, acrimonious, and politicized.

Crime will continue to be a factor affecting structure and design. Most attempts to deal with crime are now ad hoc and add on. On the other hand, much of what is now the crime-prone population will shrink in numbers with shifting age cohort patterns. There will be fewer people in the crime-prone age bracket, age 16 to 24, and hence, an automatic decrease in crime and its perceived risks. Likely to loom larger in the future, however, is political terrorism. We already see that the political factions of the Middle East, Latin America, and even southern Europe are now prepared to wage terrorist wars wherever an attractive target appears. The United States has not been immune to this, and its relative resistance is likely to decline. Increasing design attention to guarding against terrorism need not imply fortress structures. It may imply greater attention to information, identity cards, checkpoints, security passes, thumbprints, and all the accoutrements of what Bertram Gross calls a "friendly Fascism."

Although people are willing to take recreational risk they have a strong aversion to collective risk in transportation, housing, communication, and so on. Risk aversion will be reflected in closer attention to the problem of urban fires. Scores of fire chiefs suggest that there is a widespread belief that there will be one or more highrise holocausts before the supine position of public officials, the greed of builders, and the indifference of money-lenders are overcome in more rational and fire-resistant structures. Spanning the problem of fire and indoor pollution is the problem of toxicants produced by fire. The recent case of a state building in Binghamton, New York, incapacitated by an indoor fire that polluted the whole building with PCB-laden smoke is exemplary of short-term design failures. There will be design and structural solutions.

Floods, earthquakes, storm surges, ground subsidence, and other natural hazards will be a more important factor in design. The Corps of Engineers' long-term flood control program has been a short-term success and a long-term disaster. This resulted from two factors. First was the false confidence about security on the flood plains. This led to over development which, in many cases, worsens the flood damage potential. Second, the civil works have often worsened the flood potential of the river itself. In many river basins there are more people and more property at greater risk than thirty or forty years ago. We can expect that the design implications of the national flood insurance program will become a reality.

There is also the inevitable if not imminent earthquake disaster in Los Angeles or San Francisco that might be preceded by a worse disaster in Missouri or the Ohio Valley. Ultimately, an integrated, so-called all-hazards design and planning approach to the built environment will prevail, including technological and industrial hazards.

CHANGES DUE TO
TECHNOLOGICAL DISPLACEMENT

Rush hour may become a thing of the past in many cities as work-distributed centers and round-the-clock work become integrated with work at home and flextime schedules. American is rapidly moving toward the 24-hour-day. One can already see this smearing out of circadian rhythms in all-night talk shows, food stores, and movies, night shifts in banks and computer centers, and in robotized three-shift factories. As technology displaces some hand labor in municipal services, there will be more automated trash pickup, automated sorting centers, and perhaps even less-noisy municipal vehicles. The technology of genetic manipulation may affect cities in making it practical to rapidly reforest urban streets, parks, and woodlands with trees that will mature faster and be more disease-resistant. Electronic mail, electronic funds transfer, and other kinds of electronic payment systems may reduce the amount and type of mail, although electronic mail is not likely to reduce the pattern of magazine and newspaper deliveries in the immediate future.

Cheap transportation is a thing of the past by the standards of the late sixties and early seventies. However, as more people around the world are acquiring an absolute and relative degree of prosperity, tourism will increase in large and small cities. The shifting pattern of oil, as already mentioned, will not only reduce the amount of natural gas and petroleum fuel oil used in the long pull and affect that delivery network, it will stimulate the use of solar collectors, windmills, solaria, greenhouses, and even hydroponics. Large and small vegetable gardens may become popular in cities under (mushrooms), at, and above ground.

TECHNOLOGIES INTERACTING
WITH OTHER TRENDS

It is unrealistic to talk about any technological developments without mentioning the trends they interact with. Consequently, much of the material above is pallid because it is not folded into other developments or because it takes those trends as implicit. Let us now turn briefly to some other trends and see how they may interact with the physical technologies of the future city.

The aging of the population will increase the demand for smaller facilities. On the other hand, it will stimulate the growth of geriatric communities and continue the migration to warm climates. Insofar as sturdy, large houses exist, aging will promote a resurgence of group living and rooming and boarding houses. This, of course, will fit with the development of the single-parent family, of the deferred marriage on the part of young people, and so on; so several forces operate toward both smaller, modular facilities and larger, multioccupancy innovative arrangements.

Going hand in hand with, but substantially distinct from kinetic structures are computer-assisted design (CAD) and computer-assisted manufacture (CAM). The ability to design devices on a screen rapidly, quickly, and reliably should open up structural design possibilities and stimulate a tremendous amount of movement in two perhaps contradictory directions. One is toward innovation because designers will be able to try and test fairly cheaply using CAD. The other direction is toward cost cutting and economizing by being able to trim, trim, trim. Both of these trends will occur and will in the short run result in more sleazy structures. A convenient analog is the terrible plastic products of 1950-1965 when cost, not quality, dominated.

But the real future hope is in new designs and new materials assisted by CAD and linked to CAM. CAM is a factory process that will move us more rapidly toward what has been a growing potential for several decades, the prefabricated, premanufactured structure. Various forms of modular structures, prebuilt units, and prebuilt elements will come on to the construction site in the future. Complementing that, of course, are the major developments in robotics that will permit much more effective operation in dangerous construction tasks, simply because the robot will be bigger, stronger, and more responsive than present equipment with human operators. Robots will give a degree of sensitivity and control to heavy-duty equipment that now can only be achieved through human senses.

In general, future urban structures will be cybernetic in the sense that homeostatic mechanisms will be built into them from the beginning—control of heat, light, ventilation, maintenance, and so forth.

A quite different development reflecting the need to deal with energy problems will be a more widespread use of underground space. The abandoned mines of Kansas and Missouri, which are almost all cut in rectangles anyway, are extremely attractive as underground manufacturing, sales, storage, and even living space. More and more public buildings will be going down as well as up to control heating and cooling energy costs.

Materials technology will affect design of the cityscape through at least four areas of development. First are new polymers reflecting the development of polymer alloys. Chemists have worked out the full range of elementary polymers available from petrochemicals. Now the movement is toward analogs of alloys, materials based upon intimate molecular-level combinations of polymers to create better characteristics. Second, ceramics will undoubtedly become more important in construction. Composites are a trend-setting development. The most familiar composite is fiberglass. Radically new physical characteristics can be designed and built into composite materials. Carbon fibers embedded in epoxy plastics will create skins of incredible strength and resistance like those now used in high performance aircraft. Wood may enjoy a renaissance as a construction material when genetic manipulation begins to modify the characteristics of wood.

What transportation technology does to the city is an open question. The reality of moving sidewalks and the trend to various kinds of transportation amenities will continue to develop. These, coupled with energy shortages will lead to the use of passageways linking buildings, as is now common in Canada, Minnesota, and a number of cities elsewhere in the Frostbelt. Above-ground transportation facilities will be augmented by new kinds of people movers, moving sidewalks, and moving chairs. Other cities will enjoy the benefits of canals. The development of interchangeable modularized packaging makes it economically attractive to ship from land to water and back. Hydrofoil bases may become more popular on rivers and in the coastal zones.

What happens with the automobile and mass transit is, of course, the widest of open questions.

Women's massive entry into the work force will have major technological implications. It will stimulate the demand for services and prepackaging and the reduction of gratutitous and unnecessary household tasks. The demands for technology will be fed by the larger discretionary income and the more expensive lifestyle of the dual or multiincome household. One can expect a great explosion of convenience technology coupled with monitoring and regulatory technologies for the household. Going along with this will be increasing amounts of recreation. Not all of it will go to the work-study center, of course, but a lot of it will.

Industrial migration and the human migration accompanying it will create new towns and boom towns in the South and the Southwest. It will also create a new kind of self-destructing community—one of planned, long (20-50 years), but finite life. The phenomenon now common in Vermont will occur in many places throughout the country, in which the new immigrants move to the rural or semi-rural area to seek the amenities that make that kind of life pleasant. After they have been there awhile they insist on the introduction of the urban amenities that they have not psychologically abandoned. So the urbanization of new communities, small towns, and small cities will undoubtedly be a factor in the future.

International travel and high levels of education in American society will continue to upgrade taste and stimulate a desire for diversity in decor, design, and furnishings. One can see this in the restaurant business, but it will begin to permeate all other aspects of design and style. And, perhaps, the technological effects will be in interior design, while exterior designs will be more flexible. Similarly, ethnicity will have effects on technology in that the Hispanic populations will culturally influence America beyond the barrios. Most notably, new color, music, and noise will characterize many cities. Design tastes will change. One might very well see a kind of generalized hacienda becoming popular, perhaps stimulated by the general problem of security and energy conservation. Walls are on the rise.

Technology, of course, will also create the demands for continuing education, career changes, vocational changes. One might therefore see technologies of education permeating not only the work place but the home and public places such as libraries.

Let us assume that inflation and structural unemployment accompanied by relatively high taxes continue. There will be a strong shift toward a barter economy and the use of informal labor for many tasks. One might anticipate a boom in what used to be called the do-it-yourself movement. Certainly there will be a resurgence in people actively participating in the constructing of their own urban world with their own, hired, and traded labor.

The trend toward improved quality of life and greater attention to the human body as a healthy organism will cause our urban landscape to be laced with outdoor recreational facilities. Jogging is still at a relatively primitive stage compared to what one might anticipate in the systemic outdoor athletic facilities of the future. Although the odds are unclear, there may be a resurgence in bicycles and human-operated vehicles with street layouts, if not whole grids, dedicated to them.

Turning to history, the search for roots, characteristic of a mature society, is reflected in the growing concern for historic preservation. We see that initially in preserving individual houses. In the future one can anticipate neighborhoods, sectors, industrial facilities, and whole cities being candidates for such preservation. Within the urban landscape there would be subsidiary structures and, perhaps, even superficiallly subsidiary lifestyles and functions reflecting that preservation. Some of that will be in the form of living museums. Much of it will be superficial preservation of residential and commercial structures. How preservation affects office buildings and other kinds of public places remains to be seen. What, for example, will be the fifty-year future of the Empire State building?

IDIOSYNCRATIC AND NONCE
TECHNOLOGICAL DEVELOPMENTS

What in the future will be equivalent to Los Alamos, Cape Canaveral, and the boom town? Certainly, new cities will grow up and reflect the mining and mineral extraction of the West, and one can anticipate that the boom towns may be more intelligently designed for finite lifetimes and that cities will literally come and grow and go, planned with high levels of amenity but designed to pass off the scene without trauma. Nonce events such as the inevitable submergence of Houston below sea level, if current levels of water extraction continue, will lead to the possibility of a thousand-mile dike or dam to fend off the waters.

THE INSTITUTIONALIZATION OF TECHNOLOGY

What will be the fast-food chain and the movie houses of the future? The most certain forms of derivative technological developments will frame themselves around information technologies and new demands for human services. Technologically mediated information and advice services of every sort are proliferating. But, with the implicit capabilities of telecommunications networking for the formation of political and special interest groups, organization mechanisms will grow up for mediating those new interests. You may be able to locate and communicate in real time through a mediated service with everyone else in the country interested in ninth-century Chinese pottery, or you may be able to find thirty people nationwide hooked on Andean dances. As more practical matters, people in specific occupations and crafts can come together. The Oklahoma educational system already has the capability to deliver simultaneously educational material to people at their work sites 500 miles apart. Services no longer need be limited to the work site, nor need they be limited to education. All kinds of mind-on and hands-on consultation services will grow up. Psychiatric counseling by any of the fifty schools now operating will become routine. Various kinds of diagnostic health and human, social, and sexual services will be technologically augmented.

Analytical services and information-collection services will grow up. As we now understand it, some of the biggest markets for Radio Shack microprocessors are in Iowa and the Midwest where farmers are using micros as a way of achieving some independence from the county agent, the big buyer and seller, and the weather forecaster. They are beginning to move to more independence and to arrange their technoeconomic affairs more to taste. Services will grow to feed that new development.

Other derivative technological developments from CAD, CAM, and robotics will be an explosion of personalized artifacts. There is no reason why every carpet or rug or every garment could not be woven completely to order with whatever message, pattern, or image one might want at a relatively small increase in cost. One will be able to approach the design and structure of the physical world in similar ways. Second-hand markets and after-markets will become an important derivative service of robot technology. Video auctions of a $250,000 robot earth mover involving bidders simultaneously participating over a 10,000 square mile area will be commonplace.

As health technology becomes more sophisticated and cheaper, and as the costs of the medical apparatus continue to move through the roof, distributed health centers going well beyond the traditional YMCA-type training facility will flourish. There, people will be able to do 95% of a physical maintenance program by themselves with the minimum assistance of a clerk, paramedic, or a video callup system.

These are just some of the things that technology has in store for us. Most are inevitable, some are plausible, and only a few are merely speculative. The appended scenario covers much the same ground in a different style.

INTEROFFICE MEMO

A SCENARIO

Thursday afternoon late, Will had enough. It wasn't just that he'd had to work overtime for the last three days. And it wasn't that his boss kept damning him with faint praise. Will knew as well as anyone else that a good systems design should be elegant, parsimonious, and all that sort of thing. But he was only two years out of school and designing whole software packages was a tough job and they ought to be pleased he was doing as well as he was. After all, he was the youngest in the department. Anyway, his boss, Sylvia, thought the job he just finished was "O.K." She never had sent forward the report he'd done on programming for management of multilingual texts. It wouldn't even have been so bad if he hadn't learned that Ron, who had joined the company at the same time, was making $2000 more than he.

At any rate, Will wrote out a masterpiece of a coolly dignified resignation letter and punched the send button. It was addressed to Sylvia with a copy to

personnel and a hard copy for himself. Then he went over to see his friend
Alan and persuaded Alan to go out with him. Over a big pizza and two pit-
chers of Iron City, Alan was quite sympathetic. He encouraged Will to apply
at the Fire Control Center where he worked. They were looking for someone
to overhaul their models for allocating fire station equipment and also for
help in redesigning their alarm and locator package. As the population
shifted among parts of the district and people submitted changes in their
room-occupation charts, information kept backing up. And then not all
dwellings had automatic sprinkler systems, and there were more and more
electrical fires. It was all very complex. You couldn't send out a full set of
trucks, robots, crews, and different types of fire suppressants every time
somebody's alarm went off. They needed a good system overhaul. It all
sounded rather too life-and-death to Will. He would have preferred something
more like automatic translation work.

Next morning Will found himself at work before he realized that he really
didn't need to go there anymore. Nobody said anything out of the ordinary to
him. He settled comfortably into his purple wraparound terminal chair with a
hot cup of coffee. He logged in on his terminal, picked off the morning's
mail, and stared into space. After a while he reran the mail and found the
latest issue of Datamation. He advanced to the professional vacancy ads. The
first one that caught his attention was one of the local steel mills. It looked
like they had just gotten a new, upgraded robotic fork lift for shifting stacks
of plate, and they needed to reanalyze their whole production system to make
optimal use of it. A possibility, but the pay wasn't any better than he was
getting at Westinghouse. Had gotten, that is. Other ads were for faraway
places. Should he think about leaving Pittsburgh? Leave all his friends and
family? Maybe for Hawaii. Was there anything in Hawaii? Probably not.
What about the university?

Will called up the local employment exchange. A face flashed on. She was
really lovely. Wow. Could she help him? Yes, indeed. He blushed, and explained that he was interested in mid-level systems analysis work, perhaps at
one of the universities. Several listings appeared on the screen. No, they were
all too humdrum. The Face returned. What was wrong? Nothing, really, but
didn't she have anything more interesting. More listings. The Steelers! Work
for the Steelers? It was to maintain their scouting system. Good pay. Did he
have enough football experience? He had been on the taxi squad at Carnegie-
Mellon his first year. Will decided it was a long shot. As he gazed at the image
it faded. What happened? He called back and asked the Face, "What happened to the Steelers ad?" She said if it vanished the job must have been
removed from the file just then. It must be filled.

Then Will got a call from the Energy Office. His request for extra energy
credits would cost him $200 this month. (Damned Fiat!) He punched in his
bank number. Only $68.30 left. And where would his trip to Westworld come
from? And the 25% rent increase? Slowly he decided it had all been a
mistake.

Reluctantly, Will called John Little, the personnel director. John was busy.
Sylvia called him. Was he sick? Why had he not even called up his work

folder yet? Was she playing games with him? Did she know? He couldn't tell from her expression. He said he had a problem and could he talk it over with her. Yes, right after lunch. She had a lunch date with her boyfriend who was on assignment in Alaska and they were supposed to dial up at 1:00. It was breakfast for him. Suddenly, John returned his call. Will asked for a personal meeting, and it was arranged quickly. John had had a cancellation—some interview call to Marseilles and the translator was down.

John had one of the few offices with real walls in the building. It was good to have some privacy. Will explained that he had sent an unfortunate letter, and wanted to cancel it. John was very sympathetic. "Gee, Will, you're one of our best young people. I'm really glad you've changed your mind. Maybe you'd like to join one of our Newer Employees Groups. They meet every Monday at noon on the screen with a counselor. It seems to be really helpful." Yes, well, Will would indeed like to have an opportunity to compare his experiences with others, but was it too late? As a good systems man Will understood how hard it would be to undo his resignation once it got into the system. John called his assistant. Yes, the resignation had been processed. They stared at one another.

Will spent the afternoon tracing the news and sending cancel messages. First there was the letter to Sylvia. He sent her a recall. He'd see her soon anyway. What if she decided she was better off without him? He couldn't take time now to think about that. He tried to query her INBOX, but he didn't have the code. Then there was the personnel office. His file had been pulled and sent to Harry and an exit interview scheduled. He sent a message to Harry, and formatted another for Harry to use in sending his file back. Then there was payroll, the insurance companies, and his bank. The insurance companies were a problem; he couldn't get a confirmation from them. Who else? There was the other side of personnel. As it turned out they were already formatting an advertisement for his replacement. And there was security. They were in the process of putting through cancel orders on his code. He caught them just in time. They would also need to speak with Sylvia before he could be reinstated.

That worked out just fine, because it was time to speak with Sylvia anyway. She was preoccupied and hadn't even read his letter yet. They had a long talk about his work and their relationship. She gave Will a new assignment to design an elementary school learning package that could be attached to any of a number of standard commercial interactive programs and detect dyslexic response patterns, reporting to all relevant offices. Sylvia also reported that their unit had been chosen to participate in a study of time budgeting which Westinghouse had asked a consulting group to perform. She asked Will to record his activities for the next day in which he had a work period on a 15-minute basis.

Will felt relieved that his mistake had been undone and pleased with his new assignment. He was still vaguely unhappy with his job, but at least he had a job. Resolving to do better about saving money, he hung it up for the day, signed off, and went home. After a little racquetball and some swimming at the spa he left a request for a reanalysis of his exercise program and went home. Tomorrow was the last scheduled work period until late next week. Before bed he read the instructions for his daily activity record form. Will slept rather late on Saturday morning. It was 8:30 before he guided his small electric car into the automatic highway. No more Fiat until payday.

As soon as he got to work Will signed on and called the library. He didn't know anything about dyslexia. The librarian appeared on his screen. After a few minutes' conversation they agreed that he needed to know more about symptoms of dyslexia, pattern recognition, the structure of elementary school learning programs and systems, and elementary school responses to dyslexia, for starters. Together he and the librarian examined the thesauri for a number of bibliographic retrieval systems and decided on the best search request patterns.

Will spent the morning with the retrieval systems, examining titles and abstracts, calling up sample documents, refining his requests, and learning about the shape of the literature. After lunch he talked with the librarian again. Finally, he requested printouts of a number of references and abstracts, ordered up the full text of everything written by Wilson Shepherd, who apparently was the leading expert in dyslexia diagnosis, and asked for the leading text on pattern recognition to take home with him. Then he took a break while he waited for the university library to send over a computer output microfiche of the text. While he waited he filled out his time study form. Will lied about the rest of the afternoon.

11

Toward the Androgynous City

SUSAN SAEGERT

☐ THE NOTION OF AN ANDROGYNOUS CITY needs some explanation. For the idea to make sense, it must be prefaced by reference to a previous article in which I have argued that cities in this century have been identified primarily with masculine pursuits and imagery: the city as aggressive, energetic, dangerous, full of powerful forces, reckless of human life (Saegert, 1980). The idea of the city as masculine has been accompanied by policies, planning, and, more important, development that has increasingly segregated stereotypically feminine activities and concerns related to domestic and private life from places of political and economic power. More and more, the official place for female activities, the home, has become isolated in suburban repose and homogeneous residential areas.

Two psychologists, Sandra and Darrell Bem (1970), introduced the concept of psychological androgyny as a corrective to what they identified as the unconscious ideology of sexism underlyng dominant conceptions of personality. They argued that the fully developed personality could combine attributes like assertiveness and independence traditionally associated with male character with positive female attributes like nurturance and empathy. I am following their example by suggesting that the policies and actions that have led to the urban forms we have today have been unconsciously biased. Masculine values and pursuits have had a greater role in shaping the cities in which we live and work. In addition, the form of these cities assumes a sexual division of labor in which men work outside the home and are intimately engaged in the public and cultural life of the city whereas women stay at home and tend to the needs of their family. In turn, the physical form of the environment supports the continued gender differentiation of roles within the home.

Urban metropolitan development reifies the idea of the typical household as composed of a working husband, a nonworking wife, and children. Those living in this reification find the form fits most easily when they themselves approach conformity to the stereotype. For example, Rothblatt et al. (1979) report that married nonworking women were the ones most satisfied with their housing in largely suburban San Jose, California. Women and men living in suburban homes conform more closely to traditional standards of gender-differentiated roles and divided household tasks in a more traditionally gender stereotyped way (Saegert, 1980).

This chapter will attempt a fuller examination of relationships among urban forms, sexually based division of labor, and the different life experiences of women and men.

SEX-SEGREGATED WORLDS
AND DEMOGRAPHIC TRENDS

A contradiction has come into existence in the last three decades between some aspects of metropolitan form and shifting demographic trends. Since 1950 central cities have steadily lost population as suburban rings have grown. Between 1950 and 1970 suburban growth has exceeded central city growth by a factor of between four and five (Rothblatt et al., 1979). This rapid explosion of suburban residents came mainly from the parents of the baby boom, young couples heading off to FHA-financed housing on government-supported roads to establish their own little nuclear families. Wives were to stay at home and lead the life described by Betty Friedan in *The Feminine Mystique*.

While suburbs have been developing apace, the prevalence of the kinds of households they appeal to most has been declining. Three factors have decreased the fit between early conceptions of suburban housing and population trends: (1) increased female employment, especially of married women; (2) increased numbers of single people living alone; and (3) increased frequency of female-headed households. By 1978 almost 60% of all women were in the paid labor forces, including 57.9% of those living with their husbands and with children. Since 1947 women's labor force participation has increased by 18%, and that of married women by 28% (Smith, 1979). The same source predicts that female labor force participation will rise to 68.3% by 1990. By then, 70.1% of married women with children older than six and 55.3% of those with children under six are expected to be in the labor force.

During the period of rapid suburbanization, the incidence of divorce also rose steadily. This upward trend has been relatively constant since about 1920, except for an immediate post-war peak in both marriages and divorces. Between 1967 and 1977 the rate of divorce per 1000 population increased from 2.5 to 5.0 (Smith, 1979).

While sophisticated analysis of 1980 Census data is not yet possible, preliminary data indicate that the majority of adults are now married. However, almost 40% of women and 33% of men are either single, separated, divorced, or widowed. Ross and Sawhill estimated in 1975 that the fastest-growing type of household was female-headed. Today eight million households are headed by women, including over five million with children under eighteen. These female-headed households make up 14% of all families.

Single people living alone increased 65% between 1970 and 1978, from 11 million to 17 million. These single people really form two distinct groups; a burgeoning proportion of the under-30 population who are not marrying and a growing number of elderly widows. Of men 25 to 29 30% remained unmarried in 1979 while 20% of the women in that age group were unmarried. In the 20-24 cohort almost 50% of the women remained unmarried while two-thirds of the men were not married in 1979. Despite the increase in the young singles population, women over 65 years of age made up over half of all adults living alone.

ANDROGYNOUS IDEALS

These trends present challenges to existing settlement patterns and policies in that more women combine domestic and work responsibilities while frequently continuing to care for children. Large numbers of them must do so without the aid of a mate. More single people seek full lives without an opposite sex partner. Yet demographic changes are but one force in an impetus toward an androgynous city.

The idea of an androgynous city involves a change in peoples' transactions with the city that encompasses both their values and emotional experiences as well as their functional relationships. Because we are focusing on the androgynous city rather than the androgynous person, these changes could occur in a number of ways.

The key element of such a city is that transactions with the environment that have been characteristic of, or more available to, one sex or the other would now be available to women and men alike. In principle, such a city could exist and yet women and men would choose to experience it in different ways. In that case, the major difference between the androgynous city and those we have today might be that the pursuits of women would be given a more prominent place in cities. Reproduction and domestic life would command resources and status equal to those accorded male pursuits. This kind of model of the androgynous city was advanced in Skar's (1981) account of the Quechua Indians of Peru.

An anthropologist may approach this task of defining the androgynous city by looking to lost or distant societies, a science fiction writer by drawing out new visions. I look for the androgynous city from the perspective of my training as a social psychologist, from my research on the meaning and use of environments for men and women, girls and boys in metropolitan areas of the United States in the 1970s, and from my work with architects and urban planners to take a next step toward a new urban form that would better serve people in this era. These roots are very much anchored in a particular time and place.

First, empirical research tells us how the sexes differ and sometimes the conditions that magnify or minimize these differences. It can not reveal to us that which does not exist. Second, physical and social planning express conceptions of social goals held by decision makers, professionals, and various constituencies. The definition and pursuit of these goals arises from particular economic and social processes. The analysis of these aspects is beyond the scope of this chapter. Undoubtedly, my sense of potential steps toward the creation of an androgynous city suffers from the limitations of my experience and is shaped by the apparent inefficacy of this decade's policies to equalize the conditions of women. Thus the analysis that follows is stronger on documentation of conditions as they are primarily in the United States than it is on evidence for successful solutions or descriptions of androgynous cities, if indeed they could be found.

At this particular time in American history two trends provide an impetus for the development of an androgynous city. One push in this direction arises from the changing demographic patterns already described, the second from a growing prevalence of the conceptualization of women as person. Rothman (1978) proposes that American public policies and practices have been based on four different conceptions of women that have sequentially dominated from 1870 to the present. After the Civil War a model of "virtuous womanhood" seemed to characterize American attitudes toward women, followed by the ideal of "educated motherhood." Then in the 1920s conceptions of women's rightful role took on a more sexual cast, with women's position being seen as that of "wife-companion."

Only since the 1960s has the idea of woman as "person" prevailed in public attitudes and policies.

As Rothman documents, while the idea of woman as person replaces an acceptance of the appropriateness of various inequalities in the conditions of women and men with ideals of greater equality, the policies and practices directed toward this end have not yet been notably successful. For our purposes a key element in the shift in conceptualizations of womanhood involves the change away from seeing the worlds of women and of men as inherently different. Thus an androgynous city from this point of view would be one in which the city presented equal access to resources, potential living conditions, functional convenience, and personal experiences to women and men. Before going on to speculate on the social, cultural, and physical forms of such a city, existing inequalities and divergences will be briefly reviewed.

INEQUALITY OF LIVING CONDITIONS

Two kinds of households in which traditional sexual division of activities, responsibilities, and personality types are least likely to be possible or functional share two circumstances. They are more likely to be located in cities and they tend to have lower incomes. Stein (1981) states that the majority of single adults under 35 living alone are clustered in cities, particularly New York, Chicago, Los Angeles, and San Francisco. In 1977 about 42% of all female-headed households were located in central cities whereas only about 27% of the total population resided in cities (Freeman, 1980). This percentage includes about 34% of all white female heads of households and about 64% of all black female heads of households. In 1979 11.9% of all white households were headed by women as compared to 41.7% of all black households (New York *Times,* 1982). If the city is symbolically masculine, its demographic composition is not. In the same year, a third of all central city households were maintained by a person living alone or with unrelated others. Only one in five suburban households had these characteristics.

The census data for 1980 reveal sizeable differences in income related to household composition and sex of household head. The median income for female-headed households was only $10,408 as compared to $17,519 for single men and $23,141 for married couples. The plight of female-headed households appears particularly serious. The Advisory Council on Economic Opportunity predicted that in 1980 that if current trends continue virtually all people living in poverty will be women or members of female-headed households (New York *Times,* 1981).

These figures bring us back to the link between a more androgynous city (and polity) and increased equality of condition for women. Freeman (1980) has documented the extent to which women make up a disproportionate percentage of city residents and of the poor. She links urban fiscal problems to the prevalence of low-earning women in their populations. Her policy recommendations for improving the lot of both women and cities center on insuring adequately compensated job opportunities for women in urban centers and extending operating hours of businesses, services, and institutions to facilitate personal and domestic life as well as to increase the number of jobs. She argues that the availability of inexpensive rental housing, public transportation, and proximally located services and facilities already attract women and provide a promising mileu for lifestyles free of traditional gender stereotypes.

In order to allow people to take advantage of longer hours of service, flexible time schedules or shifts would be required. Barrett's (1979) review of the relationship between work schedules can offer an alternative to part-time employment. Full-time work with flexible hours therefore might improve women's earning capacity as well as increase the potential for men's involvement in domestic tasks. A study comparing male blue-collar workers who worked four- and five-day weeks showed that those with shorter work weeks spent five times as many hours on child care, and on household chores such as gardening, errands, and repairs (Macklan, 1977). Barrett warns, however, against the possible selective introduction of flexible working hours to keep women segregated in traditionally female and low-paying occupations.

These changes would indeed contribute to the development of an androgynous city. Two caveats should be added. An androgynous city would support valued activities and experiences that are traditionally female as well as those that are traditionally male. Second, if one is concerned about the equality of women, attention must be given to the support for gainful employment and varied lifestyles offered by suburban areas and small towns. As I have discussed elsewhere (Saegert, 1980), no single community form seems particularly well-suited to provide both satisfying homes (good for childrearing, safety, high environmental quality) and good work opportunities and rewards, especially from women's point of view.

WOMEN AND MEN'S USE
OF URBAN ENVIRONMENTS

Women's ongoing responsibility for domestic life, including child care, places them in a particular relationship to the urban environment. It has been estimated that the average time spent by women in household work adds up to between forty and seventy hours a week (Hofferth and Moore, 1979). What happens then to a household with no full-time homemaker? The answer to this question of course depends on the household composition. Single people and those with no children can manage most easily; those with young children, and especially with young children and only one parent, are in the most difficult straits. Working women reportedly spend only an average of 26 hours a week in household work despite evidence that the husbands of working women do not participate in household work substantially more than those of unemployed women (Hofferth and Moore, 1979). Since the median wage for women has usually been lower than the cost of replacement by paid household help, other ways of reducing housework must be found. Four suggested options have implications for policy design and planning as well as for the temporal scheduling discussed above. Hofferth and Moore list the alternatives as (1) relying on unpaid help from spouse, relatives, and friends, (2) reducing housekeeping standards, (3) becoming more efficient at household work, and (4) only working part-time.

Numerous studies have demonstrated a link between housing location, neighborhood qualities, housing design, and women's work inside and outside the home. Saegert and Winkel (1980) found that both men and women in suburban locations reported doing more household work, although they also allocated the jobs in a more rigidly sex-typed way than their urban counterparts. While suburban women were as likely to work as urban women in this sample, they expressed less commitment to work. Michelson (1977) found that couples in which the wives worked were more likely to choose downtown apartments, that wives moving to the suburbs increased the time spent in housework and initially reported a sense of isolation after the move. Both Michelson (1977) and Macintosh et al. (1977) collected time-budget data indicating that men in suburban homes spent less time with their families, although in both studies suburban men reported having stronger commitments to family. It can be hypothesized that the truly critical issue for working couples, especially working mothers, involves the extent to which it is possible to create a system of using time and space that facilitates the performance of both domestic work and work for pay as well as leisure activities.

The studies reported are also limited in that they all sampled relatively affluent populations. A second hypothesis is that lower-income women and members of nontraditional households are more dependent on the possibility of combining all the activities of life in the particular ecology of their communities. Women and low-income people are more likely to depend on public transportation. They are less likely to be able to buy replacement services to make up for nonperformance of housework and child care. As Schorr (1966) suggested, low-income people are more subject to the opportunities and constraints of their immediate residential community since they lack many resources that would ameliorate deficiencies in the public environment. Thus, I would suggest that an androgynous city would especially benefit low-income people and women. The plight of the female-headed household epitomizes the problems that most communities present to those who must perform both traditional male and female roles.

COMMUNITY FORM AND THE COMPLETE LIFE

The types and locations of residential communities affects an individual's ability to organize all the functions of a household. Women, especially working women and single parents, share certain preferences for community form with single people (Adams, 1981) and with older people in general (Steinfield, 1982). Rothblatt et al. (1979) found that of the 825 women surveyed in San Jose, California, those who were most satisfied with their communities lived either in older, centrally located neighborhoods or in planned communities. Women living in inlying neighborhoods reported greater satisfaction with community services and transportation. They also evaluated the quality of life in these areas differently on subjective criteria. Women in inlying suburbs felt more of a sense of belonging, a fullness of life, receptivity toward the world and companionships. In this sample they were also unexpectedly more pleased with the quality of the housing itself. Older neighborhoods also were found to be more satisfactory. The environmental variables affecting housing satisfaction most strongly included (1) relatively high area density, (2) a location relatively close to the central city, (3) short length of husbands' commute (wives' commute was less predictive), (4) similarity of income and (5) education to the neighborhood average, (6) small community size, and (7) planned characteristics.

This study points to two sorts of preferences that are at odds with the popular belief that single-family housing in low-density areas is most desirable. First, both working and nonworking married and unmarried women were more satisfied with multifamily housing and with community services in multifamily housing. Married women did report more of a sense of belonging in single-family houses.

In other studies, high-density housing in areas with adequate, responsive services and easy access to cities has been found to be especially desirable for divorced women (Brown, 1978). Popenoe (1980) reached the conclusion that the social services and community planning of a typical Swedish suburb offered more support for working and less isolation for women than a comparable American suburb. These studies take us beyond a simple contrast of suburbs versus cities to a consideration of the characteristics of communities that support the full range of activities people are seeking rather than forcing them to trade off proximity to work, services, and cultural opportunities against safe, desirable housing in physically and socially comfortable neighborhoods. They begin to suggest guidelines for communities to accommodate all human activities.

Yet one aspect of Rothblatt et al.'s (1979) findings must raise questions about the likelihood that the androgynous city would indeed be a city. While suburban areas with many of the characteristics of a city, including closeness to the central business district, were positively evaluated, central city areas themselves were not. The area of satisfaction most affected by suburban versus city location involved community services. The authors suggest that desirable housing tends to be located in a small political unit because such units are more responsive to community participation in planning. While economic comparisons of the areas are not made, it may also be that the city of San Jose proper suffers from some of the fiscal shortages of other American cities in this decade. The findings that suburbanites were more satisfied with group activities and felt more receptive toward the world than urbanites may be tied to the nature of the services provided in suburban areas: parks, good schools, security, child care, transportation, and entertainment. The only area in which Rothblatt et al. found an urban advantage was in women's work satisfaction. This finding echoes a theme I have discussed in greater detail elsewhere, the superiority of cities as places to find satisfying work (Saegert, 1980).

CITIES AND MALE-BIASED PRIORITIES

Despite changing patterns of household composition and employment, cities continue to be seen as either unsupportive of domestic life or as, at best, compromise locations. The major negative features of urban residences concern fear of crime and unpleasant environmental qualities such as litter, noise, pollution and crowding, poor public schools, inadequate housing, and insufficient accessible outdoor play space and greenery.

Mackintosh (1982) found that certain design and management features of high-rise urban housing significantly affected the extent to which urban couples with children were satisfied with their housing.

In her study, couples in which both partners worked were strongly attracted to city living. For example, no full-time working women interviewed stated that their ideal home was a single-family suburban home, as compared to around a quarter of the other respondents who viewed single-family suburban housing as ideal. The design features that were most valued by couples with children committed to urban living included (1) resident-only access to on-site play facilities and landscaped areas, (2) availability of three-bedroom apartments, (3) outdoor corridors and internal play yards, and (4) two-story apartments. Kitchen size was generally seen as inadequate. The one development with cooperative ownership seemed to satisfy some of residents' desires for a sense of belonging and facilitated community and other types of cooperative activities (e.g., food and wine co-ops). The enclosed staff-supervised play facilities, and relatively responsive management practices of a second development also seemed to lead to cooperative child care arrangements and social interaction. Easy access to child facilities dramatically decreased the time parents spent supervising children outside and lowered the age at which children were allowed out alone. Single-building highrises were the least satisfactory alternatives, although people living in them tended to state that they preferred anonymous, urbane social relationships. Nonetheless, they were dissatisfied with the absence of play facilities.

Thus far we have considered the functional requirements of an androgynous city from a purely behavioral perspective. Bernard (1981) has posited a distinct female world, with a sociological function of integrating society. This world, she says, is locale or kin-based, operating on a duty and/or love ethos as compared to the economic, contractual, and universalist relations of the male world. Thus, an androgynous city would have to combine the two functionally different ways of living that perpetuate both the *integrity* (a term Bernard borrows from Kenneth Boulding) and the economy. While the premise of this chapter challenges Bernard's assumption that there will always be separate female and male worlds, a significant amount of data suggest that at present women and men evaluate their physical worlds from somewhat different perspectives reflecting different value commitments.

FEMALE AND MALE
EVALUATION OF ENVIRONMENTS

Several studies of urban housing reveal that married couples sharing the same home evaluate it differently. Makintosh (1982) found that when men talked about the advantages of urban living, they focused on proximity to work and beliefs that their housing was a "good buy,'' Women usually evaluated their urban residences positively hen they made social contacts easier, facilitated walking to services, entertainment, and so on. Men expressing a preference for urban life also mentioned the absence of stress they believed would be associated with commuting.

While 60% of the sample couples agreed on preferred location for housing, when couples disagreed men and women gave characteristically different reasons for their disagreement. When husbands were more positive about suburban living, they usually wanted to own property and have a sense of permanence. The disagreeing wives of these men usually were starting a career, and were worried about finding jobs in the suburbs and problems of child care if they had to commute to the city. These women also expressed a fear that their determination to find a job would fade if they were surrounded by suburban women. When preferences for residential location were reversed, women saw the suburbs as offering a more gracious lifestyle and more independence for children. In these cases of conflict, husbands did not like the idea of commuting and had little interest in the quality of housing.

Several studies of couples who have moved from cities to suburbs indicate that the wives often suffer a loss of companionship, cultural opportunities, and a sense of isolation. Mackintosh et al. (1978) reported that city women's attitudes toward suburban moves were complicated On the one hand they genuinely felt that their children and husbands were better off in the suburbs, but they themselves often felt cut off, in danger of stagnation, and sometimes seriously depressed. Generally, husbands who had moved to the suburbs found the new surroundings more relaxing and were pleased with their opportunity to become homeowners. They also valued the availability of outdoor play spaces for children. Working-class couples showed somewhat similar disparities in the effect of moves from cities to suburbs on wives and husbands (Tallman, 1969). Wives reported symptoms of depression and experiences of loss of social ties. Men, however, said that they had increased their social support because the move to the suburbs had added to their intimacy with their wives.

Young and Wilmott (1962) studied the effects of a move to suburban housing estates on working-class residents of London's East End. They demonstrated that women lost their public and social roles when they left the dense urban environment for the world of green lawns and apartments assigned by the state. Previously, women had frequently combined their domestic chores with visits and certain business activities. Kin contact had been ever present. After the move women were more isolated in their apartments and began to practice and value a more strictly personal and couple-centered lifestyle. Because of the fact that men's daily activities took them out of the home (and transportation was costly), husbands began to take over responsibility for visiting kin and even some shopping.

As Young and Wilmott point out, values can change as some opportunities are opened up in a setting and others foreclosed. Yet the central tendency in all these studies is to suggest that couples seek out and adjust to environments in such a way as to maintain traditional gender-based segregation of activities and values. Since we know little about other types of households, predictions must be precarious. But on the basis of these findings I would suggest that the combination of women's continuing responsibilities for domestic and personal life and a set of policies that rewards male activities, places economic values above those of what Barnard calls the "female world," and fosters separation of attractive home environments and centers of work and culture will militate against the construction of androgynous cities, despite increased female employment.

MERGING MALE AND FEMALE WORLDS

The forces that could turn the tide in favor of an androgynous city are the rise in single person and nontraditional households, and, possibly, greater male participation in, and valuing of, domestic, child-rearing, and socially integrative activities. In addition, those committed to female-typed values could provide examples of alternative environments and social/economic arrangements that facilitated all of life, not just the aspects deemed important by a male-dominated political and economic system (Hayden, 1980).

It would not be difficult to document the social costs, even in economic terms, of the problems encountered by nontraditional households. Numerous studies have documented the greater incidence of physical and mental illness among single people as compared to those who are married (Pearlin and Johnson, 1981; Somers, 1981). Perlin and Johnson go on to show that for the most part neither married nor unmarried people with few children, good incomes and social support suffer greatly from depression. However, increased numbers of children and decreased incomes weigh more heavily as factors leading to depression for unmarried people. Social support is more likely to exist for married people in the first place. However, the presence or absence of such support is a more crucial contributor to a good mental condition for single people. Somers (1981) convincingly demonstrates that single people, especially elderly single people, require more institutional health care and that care is more frequently provided by the state. She also states that medical evidence is growing that loneliness and absence of social support are in themselves causes of illnesses, thus further increasing the demand for medical services.

Rothblatt et al. (1979) started out to evaluate the impact of housing and community environment on women's psychological well-being and ended up concluding that marriage counseling and other supports to sustain satisfying marital relationships were more important than urban-planning policies. However, all the unmarried women in this study had at least one child in elementary school living at home. In addition, the area studied was quite suburban in character. Education was a second factor related to psychological well-being, suggesting that policies supporting continuing education of women can have a beneficial effect. However, the third and fourth life-cycle variables relating to well-being return to the problem of creating more fully human definitions of life and community. Older women were more likely to score low on psychological well-being, despite the other trend for mothers of older children to report better psychological well-being. Here we are reminded once again of the burdens mothers feel in raising children and of the impoverished view of human development inherent in a conception of womanhood centered around youth and sexual desirability. While little evidence exists directly bearing on the interrelationship of ecological factors and well-being, Maas and Kuyper (1974) do report that women in isolated suburbs were more likely to be withdrawn and depressed in old age. Women living in communities with occupational and social activities readily accessible appeared the best off. Environmental conditions had no effect on the personality development and well-being of men in the study.

These issues again suggest that a less-stereotyped view of men and women and the devotion of community resources to what have been considered domestic and social goals would be likely to improve the well-being of the citizens, at least of the female citizens, of such a city and polity.

CHILDREN AND THE ANDROGYNOUS CITY

Much of the justification for different conceptions of male and female personality and the appropriateness of gender-segregated activities derives from women's greater responsibility for child rearing. An androgynous city would therefore be required to provide social and physical environments for child-rearing that did not require the presence of an attending mother. At the least physical level, men should be encouraged and supported in nurturing children. At the most physical level, cities would have to be so organized and serviced that they did not constitute a threat to child health and development. Green space, careful planning of traffic and transportation from the perspective of both child and adult, and the avoidance of hazardous environmental pollutants would be part of an androgynous city. It could not be conceived of as a place where strong men went to contest with each other for their own and their families' daily bread; bread to be eaten elsewhere in more healthy and relaxing conditions.

Housing in this city could not be designed and built on the assumption that it would serve only the childless. This city would avoid urban conditions that have been shown to negatively affect children's social and intellectual development like residential crowding and noise. Physical designs in housing that facilitate children's access to the outdoors and sharing of child care among households would be available. These would include low-rise buildings or courtyard-type clusters around play facilities even in the tall buildings, community rooms, on-site daycare and child care personnel. Building design and management would facilitate the kind of community relationships that provide children with security and a variety of role models from whom they can learn to become socially productive adults.

The androgynous city would require changes in the traditional ways of socializing children to use their environments. Saegert and Hart (1978) review evidence showing that as girls and boys grow older, parents impose greater restrictions on girls' mobility in their environment. Boys are allowed more freedom to explore independently, and encouraged to learn to build and manipulate the physical world. Girls are kept at home learning to keep house and attending to the details of their environment (Saegert and Maltz, forthcoming). In an androgynous city boys would learn homemaking and self-maintenance skills and girls construction and way-finding. The variety of the cities' historical tradition and cultural diversity would be made available to children to suggest a multitude of ways to live.

CONCLUDING THOUGHTS

Will the city of the twenty-first century be an androgynous city? If demographic trends continue it will be, at least in some ways. But just as early refinements of the idea of an androgynous personality suggested that we must differentiate between people who have neither positive masculine nor feminine traits and those who have both, so we must be concerned that the cities of the future may serve neither men nor women. The lives of men and women may be come more similar and the environments they use less differentiated by economic versus domestic functions. Workers may plug in their home computers in isolation from social and cultural contact. Two-career families may make do in tiny apartments without play facilities rather than cope with the problems of commuting. Elderly women and men may alike be cut off from access to the places, contacts, and experiences of the broader society.

On the other hand, extrapolating linear trends has never been a very accurate form of prediction. The city of the future may not be characterized by a convergence of lifestyles of men and women. The difficulties and inadequate rewards for women fighting against policies, attitudes, and places that do not support their domestic as well as economic aspirations may induce more of us to opt for one or the other. The trend to ghettoizing the losers in this trade-off, the single mother and the elderly person, may increase.

To move toward a city that supports the full human development of women and men would require a major reordering of priorities. The unpaid and underpaid services of women that have buffered men and children from the antihuman qualities of the environment would be replaced by healthy, well-serviced cities, suburbs, towns, and countryside. The androgynous city would be a much more homelike place. The home would not be a "container for women."[1]

NOTE

1. I owe the particularly apt phrasing of these idea to, in the first case, Delores Hayden's explication of Frances Willard's vision of the city as a well-run household in Hayden (1981). In the second case, I have taken the phrase from James Duncan who entitled one of his essays "From container of women to status symbol: the impact of social structure on the meaning of the house," in Duncan (1982).

REFERENCES

ADAMS, M. (1981) "Living singly," in P. J. Stein (ed.) Single Life: Unmarried Adults in Social Context. New York: St. Martin.

BARRETT, N. S. (1979) "Women in the job market: occupations, earnings and career options," in R. E. Smith (ed.) The Subtle Revolution. Washington, DC: The Urban Institute.

BEM, S. L. and D. J. BEM (1970) "Training the woman to know her place," in D. J. Bem (ed.) Beliefs, Attitudes and Human Affairs. Belmont, CA: Wadsworth.

BROWN, C. (1978) "Spatial inequality and divorced women." Presented at the annual meeting of the American Sociological Association, San Francisco, September.

DUNCAN, J. S. (1982) "From container of woman to status symbol: the impact of social structure on the meaning of the house," in J. S. Duncan (ed.) Housing and Identity. New York: Holmes & Meier.

HAYDEN, D. (1981) The Grand Domestic Revolution. Cambridge: MIT Press.

——— (1980) "What would a non-sexist city be like? Speculations on housing, urban design and human work." Signs 1980, 5 (Supplement): 167-184.

HOFFERTH, S. L. and K. A. MOORE (1979) "Women's employment and marriage," in R. E. Smith (ed.) The Subtle Revolution. Washington, DC: The Urban Institute.

MAAS, H. S. and J. M. KUPER (1974) From Thirty to Seventy. San Francisco: Jossey-Bass.

MAKLAN, D. M. (1977) "How blue-collar workers on 4-day workweeks use their time." Monthly Labor Review 100: 18-26.

MACKINTOSH, E. (1982) "The meaning and effect of highrise living for the middle income family." Ph.D. dissertation, City University of New York, Graduate Center.

——— R. OLSEN, and W. WENTWORTH (1977) The Attitudes and Experiences of the Middle-Income Family in an Urban Highrise Complex and in the Suburbs. New York: Center for Human Environments, City University of New York, Graduate Center.

MICHELSON, W. (1977) Environmental Choice, Human Behavior and Residential Satisfaction. New York: Oxford University Press.

——— (1973) The Place of Time in Longitudinal Evaluation of Spatial Structures by Women. Research Paper 63, Toronto: University of Toronto Center for Urban and Community Studies.

New York Times (1981) "One-parent families doubled in 11 years, census bureau says." October 13.

PALM, R. and A. PRED (n.d.) A Time-Geographic Perspective on Problems of Inequality for Women. University of California, Berkeley. (unpublished)

PEAR, R. (1982) "Inflation in the 70's wiped out gains from higher earnings in nation." New York Times (April 25).

PEARLIN, L. I. and J. S. JOHNSON (1981) "Marital status, life-strains and depression," in P. J. Stein (ed.) Single Life: Unmarried Adults in Social Context. New York: Martin.

POPENOE, D. (1980) The Suburban Environment: Sweden and the United States. Chicago: University of Chicago Press.

ROTHBLATT, D. N., D. GARR, and J. SPRAGIE, (1979) The Suburban Environment and Women. New York: Praeger.

ROTHMAN, S. M. (1978) Woman's Proper Place: A History of Changing Ideals and Practices, 1870 to the Present. New York: Basic Books.

SCHORR, A. L. (1966) Slums and Social Insecurity. Washington, DC: Government Printing Office.

SAEGERT, S. (1982) Downtown Denver Employees and the Market for Downtown Housing. Denver, CO: Denver Housing Authority.

——— (1980) "Masculine cities and feminine suburbs: polarized ideas, contradictory realities." Signs 5 (supplement): 93-108.

——— and R. HART (1978) "The development of environmental competence in girls and boys," in M. Salter (ed.) Play: Anthropological Perspectives. New York: Leisure Press.

——— and N. J. MALTZ (forthcoming) Girls' and Boys' Representations of Home and Neighborhood. New York: Center for Human Environments.

—————and G. WINKEL (1980) "The home: a critical problem for changing sex roles," in G. Wekerle, R. Peterson, and D. Morley (eds.) New Space for Women. Boulder, CO: Westview.

SKAR, S. (1981) "Andean women and the concept of space/time," in S. Ardener (ed.) Women and Space. London: Croom Helm.

SMITH, R. E. (1979) "The movement of women into the labor force," in R. E. Smith (ed.) The Subtle Revolution. Washington, DC: The Urban Institute.

SOMERS, A. R. (1981) "Marital status, health and the use of health services: an old relationship revisited," in P. J. Stein (ed.) Single Life: Unmarried Adults in Social Context. New York: St. Martin.

STEIN, P. J. (1981) Single Life: Unmarried Adults in Social Context. New York: St. Martin.

STEINFELD, E. (1982) "The place of old age: the meaning of housing for old people," in J. S. Duncan (ed.) Housing and Identity. New York: Holmes & Meier.

TALLMAN, I. (1969) "Working-class wives in suburbia: fulfillment or crisis?" Journal of Marriage and the Family (February): 65-72.

YOUNG, M. and P. WILMOTT (1962) Family and Kinship in East London. Baltimore, MD: Penguin.

12

Irregular Economies

JOHN P. BLAIR

☐ THE TRANSITION from an economy based on producing manufactured goods to a concentration in service activities is an important
characteristic of postindustrial society. The economic shifts will
continue to create serious employment dislocations, particularly
in the central cities of the northern industrial tier. The mainstream
economy will be unable to absorb a sufficient number of dislocated
workers because of the size and pace of the transition. Impediments to
smoothly operating formal markets will compound the adjustment
process. Irregular economic systems will evolve in response to changes
in the mainstream economy. A wider spectrum of urban residents will
be attracted to the irregular sectors and they will support a variety of
alternative, but increasingly acceptable life styles. The development of
nontraditional economic systems will instill strength in the urban
economy that would be evident to those who focus their analysis on
formal business activity.

Figure 12.1 illustrates the basic interrelationships described in the
chapter. The first section, denoted by "I" in Figure 12.1, describes
factors that are contributing to dislocations in the mainstream
economy. We show that employment dislocations will be severe in
many manufacturing cities due to both internal factors as well as
national and international changes. Many of the dislocated workers
will be unable to be reabsorbed into the mainstream economy. Impediments to a smooth readjustment are described in the second
section. Irregular economic activities will provide an outlet for some
individuals who cannot be absorbed in the mainstream economy. The
third section discusses prospects in the underground, household, and
grants economies.

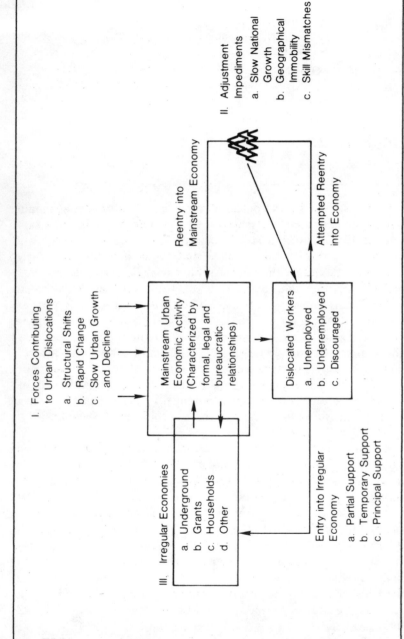

FIGURE 12.1 Urban Change and Adjustment

214

URBAN DISLOCATIONS

Urban economies are undergoing two transitions. First, the economic base of urban areas is shifting from a concentration in production of manufactured goods to a service-oriented economy. Second, the growth rate in most northern industrial areas is slow, and is declining in many central cities. These changes are symptomatic of fundamental transitions in the national economy, but the adversity of economic dislocations will continue to be felt most severely among residents of major metropolitan areas.

STRUCTURAL CHANGE

The increasing division of labor in the production of manufactured goods has contributed to the growth of urban service occupations. In order to illustrate the distinction between service as the final product and service as an occupational category, imagine the production of pencils. Milton Friedman aptly noted that no one person in the world knows how to make a pencil; pencil-making required the combined knowledge of many individuals. If, however, a single person could make a pencil, the product could be produced in a pencil factory (by production workers) and employment in ancillary service occupations, administration, purchasing, research, and marketing—would be small. As the division of labor increases, the number of service workers producing manufactured products also increases. The lawyer who has his own office but is retained by the pencil company is in a service occupation, but the final product that he helps produce is ultimately a manufactured product.

Adam Smith's famous dictum that the "division of labor is limited by the extent of the market" helps explain the growth of service occupations. Larger markets have increased the division of labor and consequently the role of service occupations. The emergence of multinational corporations and international markets will contribute to the growth of service occupations, particularly in urban areas that have adequate transportation and communications facilities to serve as international corporate headquarters. The greater the extent of the division of labor, the greater the prospects for the development of service occupations. But, internationalization of production will reduce blue-collar opportunities in cities. Likewise, increases in the ability to geographically separate management and innovation from actual production will further reduce the locational attractiveness of cities relative to small towns.

Gershung (1977) has shown that the shift toward service occupations is more significant than the growth of direct provision of personal services in England. In fact, personal services have grown no more rapidly than the economy as a whole, with the exception of education and medicine (Gershung, 1977: 111-113). Personal services are not as strong a growth sector as it might otherwise be because many personal services can be performed in the home. For reasons described later in the chapter, home production has several advantages over market transactions. Consequently, growth of personal services will be less significant than growth in the service occupation.

The Public Sector

Service employment growth in the public sector has been slow during recent years and prospects for rapid future growth are bleak. Interurban competition for industrial plants will be an impediment to raising local taxes.

The most likely areas for urban public employment cutbacks are in social services. The decline or slow growth in these areas diminishes opportunities for minorities partly because services to needy groups will deteriorate. Perhaps more importantly, public service jobs have been entry points for intergenerational advancement. Minority professionals are concentrated in a narrow range of activities such as teaching, health, social services, and recreational workers. All of these important professional tracks are closely associated with the most vulnerable public sectors (Rose, 1979: 183).

High Technology

Urban development officials are increasingly looking toward high technology as a promising growth path. The term is vague, but it suggests that urban prosperity can be achieved by following the models of the Silicon Valley or the Research Triangle. The appeal of high technology is supported by the well-known analysis of Jacobs (1970) and others who have argued that cities play a critical role in the development of new products and production processes, although the tacit assumption that new implies high technology is problematic. Furthermore, the locational orientations of vaguely defined industries is unknown, so the strategy does offer hope but not certitude. However, the high-tech development strategy is likely to be unsuccessful in many cities, partly because of the widespread pursuit of this competitive growth path. Even those cities that attract high-tech enterprises may find that the employment opportunities are related to knowledge and other service activities and not compatible with the employment needs of the local labor force.

RAPID CHANGE

Toffler (1970) expressed an important and widely felt idea when he analyzed the quickening pace of change. Rapidity of change has been particularly pronounced in economic activities. The average twenty-year-old worker can be expected to change jobs seven times in the course of his or her career, most young executives have jobs that did not exist a decade ago, the job turnover rate has increased, and employees now plan to enter job tracks, not jobs.

The ability of business to take action has increased faster than the ability of individuals and households to adjust. Mergers and acquisitions, as well as rapidly changing tastes and technology, have increased the rate of change. In general, the business sectors have been initiating change and households have been responding. The majority of structural changes will continue to result from differential rate of expansion and contraction rather than sudden births, deaths, and relocations of existing establishments. However, the overnight trauma of rapid shutdowns and discontinuous layoffs will continue to be a problem.

SLOW GROWTH AND DECLINE

Slow overall growth is possible even as the microeconomic pace of change quickens. A changing economic structure by itself would create some dislocation, but very slow and negative macroeconomic growth in many cities will greatly compound the problem. In many northern industrial cities service activities are becoming more important because of job losses or very slow growth in manufacturing. A comparison of major cities experiencing employment loss with cities experiencing employment gains (Urban Policy Report, 1980) showed that although service employment growth was positive for both groups, service growth was very modest in declining cities, replacing less than one-fifth of the lost manufacturing jobs. Furthermore, many cities with the highest concentrations of manufacturing employment are having the most difficulty attracting services. In contrast, the structure of the growing cities changes because of very rapid service growth.

The slow urban growth, led by declines in manufacturing, can be expected to continue:

> We can eliminate the possibilities of a substantial turnabout in the forces that have brought about a leveling off in manufacturing employment nationally and that are encouraging the move of manufacturing firms away from the central cities [Perloff, 1978].

The forces that will continue to retard urban growth, particularly central city manufacturing growth, include

(1)   Individual values and priorities (Harmon, 1977),

(2)   Resource scarcities (Miernyk, 1977),
(3)   Institutional rigidities and concern for special interests,
(4)   Manufacturing footlooseness (Perloff, 1978).

Many of these factors will also show the national growth rate but they apply with greater force to urban areas. None of the growth retardants constitute a sufficient argument to show that slow growth is inevitable. Nevertheless, the preponderance of evidence and opinion is that these factors will continue to affect economic prospects through the end of the century.

ADJUSTMENT IMPEDIMENTS

Schumpter (1947) described the process by which one sector shrinks in order to free resources necessary for the growth of another sector as "creative destruction" because dislocations (frictionally unemployed resources) are essential for economic development. Several factors, however, will make the destruction less creative by making the transition less smooth. They include (1) slow national growth, (2) skill mismatches, and (3) barriers to mobility.

SLOW NATIONAL GROWTH

The dislocations created by structural shift and slow employment growth would be easier to address with a rapidly growing national economy because jobs for the dislocated workers would be available. In a tight labor market firms in need of workers would be easy to identify and firms would be more likely to assist individuals in the transition, whether they required retraining or relocation.

The ability of the U.S. economy to maintain the high and reasonable sustained growth rate of 1950-1960 has been widely questioned. Over half of the post-World War II recessions have been in the last ten years. A Joint Economic Committee report synthesized and appraised information from 41 research papers and concluded:

> The transition to slower growth will influence all phases of the economy, affecting rates of investment, technical progress, and resource use. It will embody many potential sources of social stress [1978: 2].

SKILLS AND JOB REQUIREMENTS

Shifts in the employment base would be smoother if the skill requirements in the emerging sector were similar to the requirements in the declining sector. Unfortunately, the match between industrial and postindustrial skill requirements is poor. The postindustrial service economy will require a wider range of skills than manufacturing jobs. In particular, people processing and symbol processing will become even more important relative to the processing of things.

The postindustrial service economy will generate a bimodal set of opportunities. The managerial, technical, and professional jobs will offer higher wages and handsome benefits. Even many lower-paid information- and people-processing jobs in large bureaucratic corporations will also be attractive because of the regularity, benefits, and well-defined job ladders. However, many lower-skilled urban residents will be screened out of lowers-skilled service jobs in private sector bureaucracies because they lack the ability and social skills to take advantage of the internal job ladders in the future. Many of the retail and clerical jobs outside of private bureaucracies will be less rewarding and they will require communication and interpersonal skills beyond what is required of many production workers. Currently, retail and clerical jobs are frequently used as second- income sources for middle- and upper-income households. Another set of service jobs associated with conventional trade and tourism are a potential growth area, but are not satisfactory substitutes for manufacturing or construction. The jobs are frequently low paying, seasonal, and lack opportunity ladders. In contrast, entry-level manufacturing jobs generally pay well and have been relatively secure during the postwar era.

BARRIERS TO MOBILITY

Migration has been a traditional response to economic dislocations. Migration to cities during the era of agricultural mechanization was in the tradition of the westward migration during business slumps in the 1800s. Migration to areas of opportunity will still be a major response and will help families adjust to the changing urban economic structure. However, several impediments will hinder adjustment processes both among region and within metropolitan areas.

Some rapidly growing areas have instituted no-growth or "unwelcoming" campaigns to discourage new residents because inmigration restricts the ability of current residents to turn regional prosperity into personal advantage. Private groups and government officials in rapidly growing regions such as Oregon, northern California, and Alberta, Canada have initiated publicity campaigns to discourage new residents. Their success has been modest, however, because the principal tools for discouraging relocations are land use controls administered at the local level.

The geographic lockouts will be more effective within metropolitan areas than at the state level because cities have greater control over exclusionary zoning tools. Large lot-size requirements and excessive building standards will restrict residential growth. At the same time, both service and manufacturing are increasingly attracted to the suburbs relative to cities.

Federal programs designed to help the disadvantaged will discourage employment-related migration because they tie recipients to distressed areas. For instance, housing assistance, welfare benefits, and employment training are allocated on the basis of geographic need. When a recipient migrates to another area to seek a job the benefits are terminated. Of course, they may reapply for the same benefits in the new location, but eligibility and the extent of benefits vary to such a degree that there are high risks associated with giving up the benefits of social programs in order to relocate.

Finally, the higher rate of home ownership, particularly among lower-middle-income families, will impede migration because homeowners have significantly higher relocation costs than renters. Older workers will continue to be the least mobile.

IRREGULAR ECONOMIES

The continuing transformation towards a postindustrial economy coupled with adjustment difficulties in the mainstream economy will stimulate the development of irregular economic systems as well as the formal and informal grants economy. The theme of this chapter is that the underground, household, and grants economies will strengthen urban economies and emerge as more widely appreciated subsystems within the urban economy of the twenty-first century.

THE UNDERGROUND ECONOMY

Arrow (1974) suggested that when opportunities for gains exist, economic institutions will develop to capture these gains. Once the institutions become established they may have a growth dynamic of their own. The underground economy is already surprisingly large and many facets of its operation are becoming institutionalized. Unemployment and underemployment that result from economic transition will combine with continuing high tax rates to stimulate activity in the underground economy.

Common to the underground economy is the failure to report income to the IRS. The underground economy can be divided into two components: (1) illegal activity, and (2) legal activity that is rendered illegal because it is not reported. In the former case, income is hidden to avoid detection of its origins. In the latter case income is not reported because of the adverse financial consequences of reporting. High federal income and payroll taxes are major reasons for not reporting legally earned income. Other reasons for seeking off-the-books income is to avoid cutbacks in transfer payments that are geared to income, social security payments, and costs of other work rules.

The size of the underground economy cannot be measured accurately by the very fact that it is off-the-books. Some analysts estimate that it could be as high as 20% of the GNP (Clotfelter, 1981). Perhaps the best reflection of the size of the underground economy was summarized by Feige:

> I wish to note that I began this investigation suspecting that the irregular economy was smaller than previous estimates had suggested. I am now convinced that the irregular economy is indeed of staggering proportions and growing rapidly [1979: 5].

The size and nature of the underground economy varies among cities and is partly determined by the economic base of the mainstream economy. For example, newspaper reports indicate that Miami's underground economy is large and reflects illegal international drug dealings just as the mainstream economy is associated with the international trade and finance. Highly bureaucratic cities like Washington D.C. provide fewer opportunities for off-the-book earnings among the majority of workers because of the audit trail associated with a high proportion of income. New York's economic base is composed of a large proportion of personal service workers and small entrepreneurs, activities in which it is relatively easy to "shelter" income. In distressed blue-collar cities such as Dayton, Ohio, the irregular economy is evidenced by odd-job activities and the very substantial second-hand market.

Prospects

The underground economy will contribute to its future development as will the dislocations that will accompany the transition to a postindustrial society. As underground activities become more routine and institutionalized, a self--generating dynamic, complex supporting network will develop. The underground economy will be strengthened as opportunities for multiple exchange give more flexibility in barter. Dislocated workers who cannot get employment at the minimum wage will also be attracted to low-paying but off-the-books jobs in the underground economy as will employed workers who are on reduced hours or have experienced sporadic lay-offs.

The underground economy will also shift in composition from illegal activity toward nonreporting of otherwise legal activity. Demographic changes will contribute to the shift. The lower proportion of young people by the year 2000 will tend to reduce the crime rate. Furthermore, by the early twenty-first century the retired cohort will increase. Older workers may be attracted to the underground economy because their pensions will decrease with reported income. In addition, retired workers will prefer part-time or occasional work. Such jobs are often associated with the cash economy.

Policy Implications

What is the likely response to the growth of the underground economy? Recent congressional (Ways and Means Committee, 1980) and journalistic interest in the subject has emphasized the revenue loss that results from unpaid taxes. The size of the lost revenue is larger than the current deficit. If revenue loss continues as the predominant focus of concern, the response will be to close the opportunities in the underground economy. Unfortunately, such a broad reaction will neglect the positive functions of the underground economy as a safety valve to dislocations in the mainstream economy.

One approach that has been suggested to reduce unreported income is to change income tax by either adopting a value-added tax or by simply lowering the marginal tax rate. If tax evasion is possible with no risk, then most individuals will reveal their preference of no taxes rather than low taxes. So, there would still be incentives to underreport.

The increasing importance of the underground economy also has relevance for us as an urban indicator. Income and income growth, for instance, are major welfare indicators, but as the proportion of underground income increases, their ability to measure relative distress will be diminished. Similar difficulties arise in the use of unemployment and price-level indicators. The more extensive the underground economy the lower the price, and unemployment levels will be compared to the official statistics. The problem of developing indicators will be excerbated because the extent of underground activity is not equal among cities.

THE HOUSEHOLD ECONOMY

Household production refers to the organization of production within the household. Although the size and importance of household production are frequently acknowledged in economic literature, it has seldom been the center of analysis, possibly because behavior in the household sector does not reflect the behavior of economic man. Sirageldin (1969) estimated that the household economy equaled slightly less than 50% of the market economy and accounted for a higher proportion of income among low-income families and among large families. In other words, the presence of household economic activity reduced income inequalities. However, estimates of household production are inaccurate because the value of the product is not rested in the market. The pattern of the 1970s, however, was for the size of the household economy to shrink relative to the mainstream economy. The percentage of time spent on work inside the household compared to outside has declined in large part due to decreases in child care. Improvements in household technology have increased productivity, so household output has probably fallen less than work time. Pure extrapolations suggest further declines in at-home work, but several factors associated with the postindustrial transformation will strengthen the household economy.

Prospect

The household economy has been a traditional back-up that has cushioned families from dislocations in the mainstream economy. The nature of household activity will differ significantly from previous eras due to rapid postwar urbanization. Few Americans have the family farms that provided an economic base during the Depression. Consequently, alternative patterns of production organized within the household but less oriented towards agriculture will emerge by the twenty-first century.

The development of the household sector during periods of economic dislocation was documented by Morgan (1939) in classic study of how families coped during the Depression. She reported a significant increase in household production in 1933 compared to 1927. There is substantial potential to substitute household production for market activity even in urban areas. Backyard gardens, auto maintenance, and energy conservation are examples that readily come to mind. The growth of household gardens will be stimulated by slowing of productivity increases on family farms and the resulting higher prices. Output per acre in many areas is nearly equal to output on the agricultural experiment stations, so most of the benefits of current technology have been adopted. In addition, numerous household tools make substitution of household for market produce easier. Examples include auto maintenance kits, food processors, electric woodworking tools, and so forth. Burns (1975: 40) estimated that the rate of return on household capital—washing machines, lawnmowers, and so on—has been at least twice as large as the return on industrial capital (even though household capital is not recognized as investment in national income accounts). The implication is not recognized as investment in national income accounts). The implication is that investment in household capital will continue to grow because many families will invest in their household economies rather than in businesses they cannot control.

Henderson has advised "new-age" investors to concentrate on physical and human capital that can be employed in the household sector.

New Age investors are taking command of their own economic destinies by reducing their need for cash incomes, learning to coax more purchasing power out of the money they do have to spend, becoming do-it-yourself experts,, capitalizing on the untapped productive power of their homes and communities, frequently creating their own income-producing jobs, and discovering to their delight that people can be even more engrossing than products. In the argot of Wall Street, New Age investors are taking the contrary investing theory to its logical conclusion [Henderson, 1980 : 225].

Thus far, the development of the household economy has described an increasingly self-sufficient household. However, production based in the household but intended for sale in the market also has growth potential. Household-for-market production can be divided into (a) household enterprise, and (b) household-based production with a formal organization. The household-based craftsman will have two advantages over counterparts in the formal economy: (1) the ability to use the same capital, including space, for personal consumption as well as production, and (2) opportunities for off-the-books transactions and the avoidance of health, safety, and welfare regulations. The potential of home-based workers to have flexible schedules and integrated work-leisure patterns is increasingly recognized and valued. The disadvantages include reduced security and benefits, lack of customers' trust, and higher transaction costs.

Large enterprises will seek to capture some of the benefits of household production by structuring household-based jobs. The Amway-Avon techniques will continue to expand. Toffler (1980) described an ever greater potential for household-based production. Electronic communications will make home offices economically feasible. His vision of home/offices linked by wire to corporate headquarters and mainstream corporations may finance some household capital.

Policy Implications

The household economy is likely to be underrepresented in political decision-making bodies because its small scale, autonomous character, and diverse nature will make professional lobbying efforts unfeasible according to the free rider logic. The household economy is unlikely to attract the emotional or moral commitment that the profamily movement has, so it will not serve as the base of a political crusade. Powerful sectors within the market economy will continue to effect legislation and regulations in their behalf. These efforts might disadvantage the household economy (albeit unintentionally) because the household economy may be unnoticed or considered unimportant. There is currently an effort to stimulate investment through tax benefits for savers and investors. In fact, these efforts generally ignore the household economy because the inadvertant focus is only on investment in the mainstream market economy. One effect of this action is to reduce the benefits of market investment compared to household investment. Policy makers should be aware of the potential impact on the household economy because it will be underrepresented in political debates.

THE GRANTS ECONOMY

The pure grants economy contrasts with the pure exchange systems. They both are polar and ideal types. In the pure exchange system individuals exchange products of equal value. The motive of the change is to increase satisfaction, narrowly defined. Commodities including labor are normally sold to the highest bidder, although nonmonetary advantages can be conceptually embedded in price. The grants economy, on the other hand, is driven by unselfish motives like love, friendship, and responsibility. Resources are diverted from the use that represents the highest market value. Whereas the primary social function of the exchange system is efficiency, the functions of the grants economy include legitimization of relationships and integration. Boulding considers the symbolic and real sacrifices that characterize the grants economy as important threadsthat strengthen the social fabric:

> A critical question however, remained unsolved—what elements of the social systems make some conflict creative and fruitful and some conflicts damaging to all parties? I concluded that the main problem lay in what I have come to call the integrative system [1973: 6].

There are three main categories of the current grants economy: (1) intra-family grants, (2) private charity, and (3) intergovernment transfers. These categories reflect a narrow definition of the grants economy. Some writers include a variety of externalities as grants. Intra-family grants, particularly through inheritance, have been shown to explain a high degree of inequality, particularly among the wealthy (Brittain, 1978). Among families of equal wealth, however, children with lower economy potential tend to receive a larger inheritance, so in a limited sense inheritance is equalizing. Transfer payments as measured by the Department of Commerce are an important part of the urban economy, accounting for about 11% of total personal income in major metropolitan areas (U.S. Department of Commerce, 1974). Pfaff (1973), however, showed that transfers are not a major factor in reducing urban inequality.

Prospect

The transitional dislocations will place a significant strain on the governmental granting systems that developed partly in response to urbanization. Not only will unemployment rates rise and personal incomes fluctuate, but public grants will increasingly tend to be viewed as private property rights (Blair et al., 1975). Consequently, individuals will increasingly adjust their life plans to take advantage of their rights to participate in the grants economy. Partly because individuals develop life plans according to a preexisting set of grants, change in public grants criteria will be painful, difficult, and frequently unjust. The more far-sighted individual planning is, the more likely it is that public grants will deliberately support particular life styles.

The development of the household and underground economies will render reported income, the traditional criteria for assistance in the grants economy, less meaningful. Perhaps granting criteria will become more robust to account for developments in the household and underground economies. For instance, grantors may wish to consider opportunities to earn in-kind income as a criteria.

Private granting agencies will have a role in responding to changes in public grants programs. Private grants institutions will probably grow more rapidly than the public sector counterpart in the years preceding the twenty-first century but will not entirely compensate for many public cutbacks. Charitable contributions appear to be more responsive to relative rather than absolute income (Schwartz, 1976) so increasing inequality may actually stimulate granting. However, another plausible argument is that voluntary granting will decrease as the gap between income expectations and realization increases.

A major increase in the grants economy will occur through the increasing importance of inheritance. The slower the economic growth rate, the less likely that children will accumulate more than their parents, and consequently, the more important a future inheritance will be in their life planning. The development of old-age health and welfare programs have been extremely effective in preserving the accumulated wealth of elderly. Consequently, savings among the current and near future cohorts of retirees are more likely to be bequeathed. Perhaps the most important factor in increasing the size of inheritance is smaller family size. A drop in the number of siblings from four to two will essentially double the size of an expected inheritance. Larger inheritances will help cushion some families that are adversely affected by the shifts in the urban economic base.

The grants economy overlaps the exchange system in many institutions, a fact that antropologists have been more ready to recognize than economists. The grants economy will develop an institutions that combine grant and exchange systems grow. For instance, producer and consumer cooperatives have experienced modest growth recently as they have shifted to a more volunteer work force (Hall, 1980). Prospects for continuing growth are good because similartohousehold production, consumer cooperatives offer an opportunity for individuals to directly capitalize their labor. Swaney (1981) argued that pure exchange systems create a mentality in which legality becomes synonymous with prosperity. Just as most work-to-the-rule situations have counter-productive results at a micro level, a legalistic exchange system creates more externalities than a system engendered with more traditional values (including integrative grants). To the extent that this is true, cooperative and other integrative economic institutions have an important source of efficiency. Both the underground and household economies rely upon granting concepts. Henry's (1978: 107) analysis indicated that the amateur underground economy is not grounded in pure calculus of exchange but by the "etiquette of valued exchange" (Faberman and Wienstein, 1970: 450) in which reciprocity and community sharing are important motives that temper narrowly conceived self-interest.

Policy Implications

Both governmental and private granting agencies will need to recognize the emerging functions of the grants economy. Individuals will increasingly use the grants economy as a source of income to support a lifestyle outside the mainstream market economy, rather than as a form of emergency relief. Granting policy will have to address the extent that particular lifestyles contribute to larger social objectives and which lifestyles the granting agency wants to avoid supporting. The key issues in a comprehensive review will be the development of improved grants criteria.

The urban dislocations that will lead into the twenty-first century will be significant, but they will not necessarily be permanent. Granting agencies should examine their activities to determine whether they are helping individuals adjust, or whether they are contributing to an incentive system that discourages adjustment. Grant policies that tie people to particular areas and occupations that have poor growth prospects should be reexamined.

CONCLUSION AND SYNTHESIS

The inability of the mainstream economy to absorb the dislocated workers will stimulate the development of irregular economies such as the underground, household, and grants economies. Irregular activity will provide an important safety valve function.

Boulding (1973) suggested that economies are organized according to three principals: love, fear, and exchange. Given this useful organizational triology, we can interpret Polanyi (1944) as showing that economic development has followed a path away from love and fear systems, both of which require integrative social relationships, toward an exchange system that is atomistic and alienating. The great transformation from integrative to exchange systems functions best when the economy is growing. Slow urban growth and decline will cause the great transformation to sputter and, according to Polanyi, take a destructive turn toward totalitarianism. But, Polanyi's analysis can be disaggregated (Gershung, 1977). Some subsystems within an economy may move away from pure exchange mechanisms even as other sections move in that direction. This diversity may lend stability to the urban social economy during periods of stress.

The analysis in this chapter viewed the economy as consisting of overlapping and interacting subsystems. Although the mainstream economy operates principally by the exchange mechanism, other important sectors include larger doses of threats and grants. The future city economy will change in three important respects: (1) most widely noted, the activity mix will include more service occupations but not necessarily more personal service, (2) the irregular economy will assume a more important safety-valve role, and (3) the fundamental organizing motives will shift among economic subsystems.

REFERENCES

ARROW, K. (1974) "Limited knowledge and economic analysis." American Economic Review 64 (March): 1-10.

BAUMOL, W. T. (1967) "Macro economics of unbalanced growth: anatomy of the urban crisis." Amer. Econ. Rev. 54: 414-426.

BLAIR, J. P., G. GAPPERT, and D. C. WARNER (1975) "Rethinking urban problems:inequality and the grants economy," pp. 471-504 in G. Gappert and H. Rose (eds.) The Social Economy of Cities. Beverly Hills, CA: Sage.

BOULDING, K. (1973) The Economics of Love and Fear: A Preface to Grants Economics. Belmont, CA: Wadsworth.

BRITTAIN, J. A. (1978) Inheritance and Inequality of Material Wealth. Washington, DC: Brookings Institute.

BURNS, S. (1975) The Household Economy. Boston: Beacon.

CLARK, C. (1940) The Conditions of Economic Progress. London: Macmillian.

CLOTFELTER, C. J. (1981) "Tax rate and tax evasion: analysis of micro data." Presented to the Southern Economics Association, November.

Department of Housing and Urban Development (1980) The Presidents National Urban Policy Report. Washington, DC: Government Printing Office.

——— Office of Policy Development and Research (1978) The 1978 HUD Survey on the Quality of Community Life. Washington, DC: author.

FABERMAN, H. A. and E. A. WEINSTEIN (1970) "Personalizatio in lower-class consumer interaction." Social Problems17: 449-457.

FEIGE, E. L. (1979) "How big is the irregular economy?" challenge 22 (November-December): 5-13.

FORRESTER, R. (1969) Urban Dynamics. Cambridge: MIT Press.

GERSHUNG, J. I. (1977) "Post-industrial society: the myth of the service economy." Futures (April): 103-113.

HALL, B. F. (1980) "The future of consumer cooperation: the new wave." American Council on Consumer Interests Proceedings Columbia, MS: American Council on Consumer Interest: 87-89.

HARMON, W. (1977) "Contemporary social forces." The Futurist (February): 68.

HENDERSON, C. (1980) "New age investing," in F. Feather (ed.)Through The 80's: Thinking Globally, Acting Locally. Washington DC: World Future Society.

HENRY, S. (1978) The Hidden Economy. London: Martin Roberson.

HOCHMAN, H. M. and J. D. RODGERS (1973) "Utility interdependence and income transfers through charity," pp. 471-504 in K. Boulding, M. Pfaff, and A. Pfaff (eds.) Transfers in an Urbanized Economy. Belmont, CA: Wadworth.

HUMPHREY, N., G. PETERSON, and P. WILSON (1979) Capital Stock in Twenty-Eight Cities, Draft Report. Washington, DC: The Urban Institute.

JACOBS, J. (1970) The Economy of Cities. New York: Vintage.

MIERNYK, W. H. (1977) "Rising energy prices and regional economic development." Growth and Change (June).

MORGAN, W. L. (1939) The Family Meets the Depression. Minneapolis: University of Minnesota Press.

PERLOFF, S. (1978) "The central city in the postindustrial age," in C. L. Levin (ed.), The Mature Metropolis. Lexington, MA: D. C. Heath.

PFAFF, A. B. (1973) "Transfer payments to large metropolitan poverty areas: their distributive and poverty-reducing effects," pp. 93-129 in K. Boulding, M. Pfaff and A. Pfaff (eds.) Transfers in an Urbanized Economy. Belmont, CA: Wadsworth.

POLANYI, K. (1944) The Great Transformation. Boston: Beacon Press.

ROSE, H. (1978) "The diminshing urban promise: economic retirment, social policy and race," in J. P. Blair and D. Machimias (eds.) Fiscal Retrenchment and Urban Policy. Beverly Hills, CA: Sage.

SCHUMPTER, J. A. (1947) Capitalism, Socialism and Democracy. New York: Harper.

SCHWARTZ, R. A. (1976) "Personal phuilanthropic contributions." Journal of Political Economy 78 (November/December): 1264-1291.

SIRAGELDIN, I. (1969) Nonmarket Components of National Income. Ann Arbor, MI: Survey Research Center.

STERNLIEB, G. and J. W. HUGHES (1977) "The new economic geography of america." Journal of American Institute of Planners (July).

THUROW, L. C. (1980) The Zero Sum Society. New York: Penguin.

TOFFLER, A. (1980) The Third Wave. New York: Morrow.

——— (1970) Future Shock. New York: Random House.

U.S. Congress House Ways and Means Committee (1980) The Underground Economy. Washington, DC: Government Printing Office.

U.S. Congress Joint Economic Committee (1978) U.S. Long-Term Economic Growth Prospects: Entering a New Era. Washington DC: Government Printing Office.

——— (1977) The Current Fiscal Condition of Cities, A Survey of 67 and the 76 Largest Cities. 95th Congress 1st Session: July 28, 1977.

U.S. Department of Commerce (1974) Survey of Current Business. Washington, DC: Government Printing Office.

U. S. Department of Housing and Urban Development (1980) The President's National Urban Policy Report. Washington, DC: Department of Housing and Urban Development.

Part IV

Forces of Inertia:
Will Urban Change Ever Change?

☐ ALL THE BEHAVIORAL MANIFESTATIONS and consequences that will result as cities evolve through the post-affluent transition cannot be precisely projected. It can be suggested that there will be more of a divergence between the culture and the socio-economic-techno structure. Cultural style may continue to emphasize self-indulgence and hedonistic manifestations while the economy continues to require specialization, lengthy education, recurrent training, and an abstract, self-denying rationality among members of the functional elites.

But some things may, in the best and worst of our cities, stay the same. O'Brien and Clough explore the future of urban neighborhoods and report on attitudes relative to comunity and place. Fleischman reminds us about the role of quality design and the enduring elements of architecture in great cities.

Costa and Cho, borrowing from a model of city types, suggest that the enduring forces of privatism and frontierism will continue to subject American cities to the prospects of further decline. Merry reports on the realities of urban danger that will continue as long as cities remain heterogeneous and foster cultural and moral diversity. As long as sufficient crime occurs to generate anxiety, the social structure will identify "strangers" to serve as the focus of the anxiety, irrespective of the actual patterns of victimization.

Some urban conditions may never change but cities will continue as centers of innovation and creativity. Indeed, change may be seen as a source of urban stability. 231

13

The Future of Urban Neighborhoods

DAVID J. O'BRIEN
LYNN CLOUGH

☐ THE ESTABLISHMENT OF THE NATIONAL COMMISSION ON NEIGHBORHOODS in 1976 by the Carter administration marked a turning point in federal government concern with the conservation of urban neighborhoods. Prior to this, federal approaches to urban problems either were limited to specific neighborhood concerns, such as housing or transportation, or to specific target populations such as the poor (Schoenberg and Rosenbaum, 1980). The problems currently facing urban neighborhoods, however, cannot be attributed solely, or even in major part, to the impact of federal policy. Essentially, many of these problems have been generated by larger macrolevel economic and demographic trends (Downs, 1981).

In this chapter we will consider the future direction of urban neighborhoods in the United States in light of recent economic and demographic changes as well as changes in public policies with respect to neighborhood revitalization. Our central assumption is that because of their declining population and their relative loss in economic importance, cities in the northeast and north central regions of the United States will not be able to demand a great deal of attention vis-a-vis other domestic areas of concern and that policy makers with limited resources will continue to focus their energies on maintaining or upgrading working- and middle-class neighborhoods while eschewing programs that deal with poor neighborhoods.

In turn, we will propose that a major concern of policy makers will be to identify which moderate-income neighborhoods will benefit most from intervention programs. This will mean that increased attention will be given to operationally identifying aspects of "neighborhood viability." In our view the core problem in dealing with neighborhood viability is to properly understand the nature of residents' preferences and motivations with respect to participation in and attitudes about their neighborhoods.

RECENT ECONOMIC AND DEMOGRAPHIC CHANGES IN AMERICAN CITIES

Since 1970 a process has been occurring that Berry (1976) refers to as "counterurbanization," in which metropolitan areas, especially in the north central region, have been losing population to non-metropolitan areas. This has resulted in a gradual depopulation of many large central cities. Downs observes that "from 1970 to 1975, 97 of the 153 largest cities lost population—compared to only 56 that lost population in the 1960s" (1981: 104). A good deal of this shift is due to a fundamental realignment of heavy industry in the United States, with the so-called Sunbelt region experiencing a boom in the building of new modern plants to replace the outmoded factories in the declining northeast and north central areas (Lineberry and Sharkansky, 1978: 60-62).

Another dimension to the counterurbanization process, however, is that even within the northeast and north central regions, non-metropolitan areas are gaining at the expense of metropolitan ones (see Berry, 1976). Here it appears that along with the loss of huge mass production plants, any economic recovery that occurs is likely to result from more diversified "high technology" businesses, especially in the areas of research and development, which do not depend on central city location. This is seen, for example, in the emergence of the so-called industrial parks located along major interstate highways far from central cities (Lineberry and Sharkansky, 1978: 343). Therefore, in spite of whatever cultural, governmental, or other value they might possess, central cities have become less important according to strict economic criteria.

Along with the decline in the economic power of cities, there has been a substantial shift in the social and economic characteristics of the people who live in them. Despite the highly publicized gentrification process in some neighborhoods, by and large, affluent younger whites have fled the central cities in increasing numbers, leaving behind relatively larger proportions of minority group members and older whites who cannot afford to leave (Berry and Kasarda, 1977; Frey, 1979). This has resulted, then, not only in a decline in the tax base of the central cities (as well as a corresponding tax burden associated with providing services to poorer populations), but also in a reduction in the political influence of urban residents in statewide and national arenas (Lineberry and Sharkansky, 1978: 401-404).

The processes just described help to explain why interest in urban neighborhoods by both the general public and policymakers seems to have declined in recent years. Additional reasons for the shift in emphasis away from urban neighborhoods can be found in an overview of federal policies and their effects on neighborhoods as well as the experiences of various kinds of neighborhood organization efforts during the 1960s and 1970s.

FEDERAL POLICIES AND
THE RISE AND DECLINE OF
THE NEIGHBORHOOD ORGANIZATION MOVEMENT

It has been asserted that many federal programs aimed at curing the ills of urban areas have indeed been counterproductive in their influence on urban neighborhoods (Goetze, 1979; Goetze and Colton, 1980; Schoenberg and Rosenbaum, 1980). Schoenberg and Rosenbaum, for example, note that agencies created during the Depression to encourage the flow of credit by providing protection to financial institutions and establishing insurance for long-term loans did stimulate construction, but primarily through increasing the number of new homes in the suburbs that could be purchased by middle-income families who, of course, fled the cities. Alternatively, a larger proportion of those remaining in the cities were poor persons who could not afford decent housing in the private market and came to depend more upon public housing. Compounding the effect of a dispersal of population to the suburbs was the federal government's involvement in the construction of highways, making it posible for people to live further away from their work. An additional negative consequence was the destruction of the physical environment of urban neighborhoods as new federally sponsored highways wound their way through the city (Schoenberg and Rosenbaum, 1980).

Further deterioration of urban neighborhoods has been attributed to the federal urban renewal programs (Sanders, 1980; Schoenberg and Rosenbaum, 1980). Sanders observes that "enactment of the Housing Act of 1949 provided a strong symbol of national concern with the problems of slums and blight. However, the original program of slum clearance and redevelopment was not designed to aid the slum dweller directly but to alter the shape of urban land uses" (1980: 104). Frequently, rather than eliminate slums these programs merely pushed poor people into adjacent slum neighborhoods, thereby greatly increasing the density of the latter and further contributing to their decline as well as worsening the plight of the poor.

The 1960s will be remembered for a number of important social movements and conflicts, including the Vietnam War protests, the civil rights movement, and the youth counterculture, but it will also be recalled as a time of the rediscovery of the poor and of the local neighborhood. Conservatives in the United States always had defined the neighborhood as being of great value, but during the 1960s persons who traditionally had defined themselves as liberals or radicals came to see new hope in the "grassroots." The Economic Opportunity Act of 1964, which created the War on Poverty Program, was the first form of federal urban policy that specifically focused on the poor as a target population. The Demonstration Cities and Metropolitan Development Act of 1966 (Model Cities Program) was intended to test the effectiveness of a "concerted attack on social and economic problems as well as physical decay" (Schoenberg and Rosenbaum, 1980: 22; Kaplan et al, 1970).

The romance and fascination with urban neighborhoods began to wane, however, during the latter part of the 1960s and into the 1970s as failures occurred in the ambitious programs targeted to revitalize poor neighborhoods. The much-heralded "protest social action" approach of Saul Alinsky and the New Left turned out to be much less effective as adversaries learned ways to defuse incipient confrontations (O'Brien, 1975: Chapter 4). Involvement of the poor through "maximum feasible participation" in the government-sponsored War on Poverty program produced a brief period of intense conflict between established political authorities and "spokesmen" for the poor who tried to use the programs as a means to develop a countervailing base of power in local politics. Eventually the established authorities won out and the political organizing dimension of the Community Action Programs was substantially curtailed. Moreover, in both the private and public efforts, little attention was given to basic problems of creating incentives to induce the poor to make contributions to collective organizing efforts. Thus, not surprisingly, the vast majority of persons in poor neighborhoods remained uninterested in neighborhood organization programs (O'Brien, 1975: Chapters 4-5).

Most important, the difficulties experienced by neighborhood organizers during the 1960s and early 1970s highlight an inherent difficulty in grassroots organizing at that point in American history. Indigenous neighborhood organizers were in a relatively weak position vis-a-vis public service bureaucracies and other service producers in their efforts to generate inducements with which to get people to support neighborhood collective efforts. This was in marked contrast to the situation faced by the political entrepreneurs in the Eurpoean immigrant enclaves of the nineteenth and early twentieth centuries when the level of service delivery by public producers was quite low and thus the immigrant organizers would attract support for their organizational efforts by offering people selective benefits in the form of social services (e.g., help in finding jobs, charity assistance, etc.) that they could not obtain elsewhere (O'Brien, 1975: 35-42, 189-194). In a more general sense, the neighborhood has become less important to individuals as citywide, statewide, or nationwide service producers have taken over functions that at one time were provided by the local area. Thus, despite all of the rhetoric of the 1960s and early 1970s, the neighborhood remains for many Americans a "community of limited liability" (Fischer et al., 1977; Janowitz, 1967).

The failure of the neighborhood organizing efforts as well as the resurgence of economic issues undoubtedly contributed to the rather weak response of liberals to the Nixon administration's decision to scrap categorical grants dealing with poor neighborhoods, along with the entire Office of Economic Opportunity, and to replace them with the Community Development Block Grant progrant (CDBG). The CDBG guidelines not only gave more discretion to local areas with respect to where the federal money could be used (in contrast to the Johnson administration's categorical guidelines restricting application to poor areas), but actually created incentives for local planning departments to utilize funds in working- and middle-class neighborhoods where tangible revitalization results could be achieved much easier than in poor neighborhoods (Lange and O'Brien, 1978). A study done by the Brookings Institute after the first year of the block grant program showed the result was a high level of citizen participation but also a shift from a concentration on seriously deteriorating poverty neighborhoods to transitional or marginal neighborhoods (Nathan et al., 1977; Schoenberg and Rosenbaum, 1980).This change in emphasis is perhaps best articulated in the concept of "triage."

TRIAGE AND NEIGHBORHOOD VIABILITY

The concept of triage, with reference to neighborhood revitalization, was first proposed by Anthony Downs and is described in the following manner by Cassidy:

> The triage theory uses a medical analogy to explain how best to use scarce urban dollars, particularly community-development block grants. Triage is the method doctors use after a disaster. They divide the survivors into three-groups—those who will live without immediate medical attention; those who will probably die even with such care; and those who will live only if treated immediately—and then concentrate their efforts on the last group to save the most lives possible. According to Downs' analogy, the healthiest neighborhoods should get as little money as possible.... The very worst neighborhoods, the ones that in medical terms are likely to "die" no matter what is done for them, should get a small share of community development money..."the largest part of the money." says Downs, "should go into those neighborhoods where it will have the greatest effect"—the gray-area neighborhoods, which are just beginning to decline but are not beyond salvaging [1980: 285-286].

Clearly, the triage concept is biased toward investing in neighborhoods with more stable residential populations, which means that it focuses on working- and middle-class neighborhoods, while tending to avoid any effort to salvage the poorer neighborhoods that had been the focus of the War on Poverty and Model Cities programs. This, of course, raises ethical questions about different interpretations of justice (see Lange and O'Brien, 1978), but given the relatively weak political position of poor neighborhoods, the overall limits to federal expenditures in this area, and the fact that liberals have turned their attention to nonneighborhood economic issues, it seems likely that this type of policy will continue in the future.

Taking the triage principle as a working assumption, however, does not automatically produce criteria with which to decide which neighborhoods will be more or less responsive to treatment. Even if we assume that intervention strategies will focus primarily on working- and middle-class neighborhoods, the limited resources available will still require further choices with respect to the selection of target neighborhoods. In addition to the material quality of a neighborhood, there are other nonmaterial factors that lend themselves to what some writers term neighborhood "viability" (Cassidy, 1980; Goetze, 1979, 1981; Schoenberg and Rosenbaum, 1980).

The literature suggests that two aspects of neighborhood viability are especially relevant to the outcome of revitalization processes. First, Schoenberg and Rosenbaum note that "viable neighborhoods are those in which residents can control social order; residents in such areas are able to set the goals for collective neighborhood life and can implement programs to accomplish those goals" (1980: 6). The extent to which residents are able to reach consensus on neighborhood goals and "control the social order," however, would seem to rest on residents' involvement in various aspects of the associational life of the neighborhood. Knowing others in a neighborhood is likely to give residents a feeling of being in control (see Schoenberg and Rosenberg, 1980: 37) and thus the strength of informal neighboring networks in a neighborhood can be expected to have a substantial impact on its viability. More formal associational involvements, such as those in neighborhood churches, ethnic associations, and social clubs, will likely have a bearing on whether or not a neighborhood can generate the federation of groups upon which many neighborhood collective efforts are formed (see O'Brien, 1975: 30-32, 194-201).

The second, and even more important, nonmaterial aspect of neighborhood viability that is likely to have bearing on revitalization efforts is how residents' subjectively assess the quality of their local area. Several scholars (see Cassidy, 1980: 61-61, 67; Goetze, 1979, 1981) have noted that the degree to which residents' believe that a neighborhood is worth saving is bound to have an important impact on their inclination or disinclination to make material and psychological investments in their homes. When, for example, residents' perceive a neighborhood as inevitably facing decline they will be apt to see any personal investment in it as a losing proposition and thus will, by inaction, speed up the process of deterioration. Alternatively, many widely publicized revitalization efforts have succeeded in areas because residents' have defined them as having great potential sentimental or aesthetic qualities even though they were in the midst of material decline.

The important question at this juncture is: What factors are associated with variations on the two dimensions of viability described above? This was a central research concern of the Akron Neighborhood Survey.

THE AKRON NEIGHBORHOOD SURVEY

The Akron Neighborhood Survey was conducted in the summer of 1978. Akron, Ohio, with a population of 237,005 in 1980 (U.S. Bureau of the Census, 1981)[1] Our operational definition of the neighborhood was a unit called a "traffic zone" (TZ) that was used in an earlier metropolitan transportation study and for which 1970 census data was tabulated (see U.S. Department of Transportation, 1977: Chapter 2).

Since our primary concern was to examine residents' involvement in and assessment of moderate-income neighborhoods, all TZs with 1970 mean incomes less than $6,000 and more than $12,000 were eliminated from the sampling frame. A modified random sampling procedure was used to select nine TZs within three differing racial mixture types. A systematic sampling procedure was used to select residents within each of the 9 TZs. Personal interviews were conducted with a sample of 507 residents: 184 blacks and 323 whites.

NEIGHBORHOOD INVOLVEMENT

Three measures of neighborhood involvement were used in the analysis. The first was a simple cumulative index of membership in various neighborhood associations, including ethnic clubs, churches, PTAs, social clubs, and sports teams. The second item asked respondents whether they had ever "used a neighborhood organization to get fair treatment." This item was more specifically concerned with involvement in neighborhood organizations as interest groups. The last item measured the respondents' familiarity with the names of other residents in the neighborhood. This was seen as a measure of informal neighboring (for details on this procedure see Mueller et al., 1981).

Because previous research had suggested that an individual's personal resources would have an important bearing on his or her involvement in voluntary associations in general (Hausknect, 1962; Verba and Nie, 1972: Chapter 8) measures were included on respondents' occupational prestige (measured by the NORC occupational categories, Lin, 1976: Appendix B) and number of years of education completed. Individuals' racial status was ascertained for two reasons. First, a good deal of literature has shown that blacks are more likely than whites to join certain kinds of voluntary associations when socioeconomic status (SES) factors are held constant (Olsen, 1970). Second, the historical experiences of blacks and whites have been very different with respect to opportunities for movement into both neighborhood and supraneighborhood associations (Berry and Kasarda, 1977: Chapter 2) and thus it may be that individuals with different racial backgrounds will have different patterns of contemporary involvement in neighborhood associations.

Whether or not respondents were homeowners, the length of their residence in their present dwellings, and their total number of children at home were included because of the possibility that they would be associated with greater individual material and psychic investment in the neighborhood and that this, in turn, would be related to different levels of associational involvement.

The social characteristics of an individual's neighbors may be expected to have an affect on his or her involvement in the neighborhood in terms of providing or not providing a source of like-minded people one is comfortable interacting with as well as providing an aggregate level of social characteristics associated with involvement. For this reason we included two contextual variables that measured aspects of the social composition of the neighborhood. First, the proportion of blacks who were actually sampled in each neighborhood was used as a measure of racial composition. Second, the mean household job prestige scores of residents interviewed in each neighborhood were used as measures of neighborhood socioeconomic status levels. Finally, since racial composition of a neighborhood is likely to have different effects on individuals depending on their own racial status (Berry and Kasarda, 1977: Chapter 2; O'Brien 1981; Warren, 1977) an interaction term measuring the effect of individual race 9 racial composition of the neighborhood was included in the initial regression runs.

As expected, individual resources are associated with involvement in formal neighborhood associations. Both household job prestige and education are positively associated with membership in neighborhood associations and use of neighborhood organizations as interest groups. It is interesting to observe, however, that although the social composition of the neighborhood does not have a significant impact on the propensity of individuals to join neighborhood associations, there is a significant interaction between an individual's race and the racial composition of the neighborhood in the case of use of neighborhood organizations as interest groups. The unstandardized slope for neighborhood racial composition in equation 4 is -.002 for whites and .000 for blacks. It appears that when the percentage of blacks in a neighborhood increases, white residents in that neighborhood are less likely to become involved in neighborhood associations whose goals presumably are to seek revitalization of the neighborhood or to pursue some kind of issue. This interaction between individual racial status and neighborhood racial composition is consistent with other findings we will look at shortly that suggest that when the proportion of blacks in the neighborhood increases there is a corresponding decrease in the degree to which whites see that neighborhood as a place worth living in or even worth trying to save.

The most pronounced effect of the racial interaction term, however, is seen in the case of informal neighboring. Running equation 7 separated for the black and white subsamples, the co-efficients for the proportion of blacks in the neighborhood are -.009 and .007, respectively. Again, it appears that at least in this sample of moderate-income neighborhoods, blacks and whites tend to maintain traditional patterns of social isolation from one another. Finally, and not terribly surprising, is the finding that membership in neighborhood associations and use of neighborhood organizations to get fair treatment facilitates knowing other persons in the neighborhood. There is, of course, an alternative interpretation for this finding that knowing persons informally facilitates involvement in more formalized associations.

CITIZEN ASSESSMENT OF NEIGHBORHOOD QUALITY

As noted earlier, undoubtedly the most important dimension to citizens' involvement in neighborhood viability is whether or not individuals see their neighborhood as a worthwhile place in which to live. Even if a few activists become involved in neighborhood associations or interest groups and perhaps even win some highly publicized issues, it may not have any long-lasting effect on neighborhood stability if the majority of individuals in the neighborhood and prospective residents do not see it as a worthwhile place in which to live. To assess this dimension of citizen involvement, we generated a three-item index of citizens' assessment of overall neighborhood quality. The first item asked responds, "Overall, if you had to rate this neighborhood as a place to live, would you say that it is (1) poor, (2) fair, (3) good, or (4) excellent?" The second item asked, "Do you think that money put into fixing up a home in this neighborhood is a good investment?" The response alternatives were yes, no, and not sure. Finally, the third item asked, "Is this a good neighborhood for bringing up children?" The response alternatives were yes, no, and not sure.

The individual and neighborhood contextual variables as well as the various measures of neighborhood involvement described above also were used in this phase of the analysis. In addition, three variables were added to measure specific neighborhood conditions. Since crime

Table 13.1

Regression of Neighborhood Involvement on Individual and Neighborhood Characteristics (N = 449)
(Unstandardized Coefficients Are on Top; Standardized Coefficients Are on the Bottom)

	Membership in Neighborhood Associations		Use of Neighborhood Organizations as Interest Groups			Number of Persons Known in the Neighborhood		
	EQ1	EQ2	EQ3	EQ4	EQ5	EQ6	EQ7	EQ8
Household Job Prestige	.006** (.129)	.006* (.122)	.003* (.089)	.003* (.088)	.002 (.058)	.005 (.073)	.004 (.070)	.003 (.041)
Race	-.023 (-.018)	.063 (.049)	-.049 (-.060)	-.176 (-.216)	-.158 (-.194)	.091 (.053)	-.508** (-.298)	-.436** (-.255)
Education	.045* (.109)	.038* (.091)	.027* (.104)	.027* (.103)	.021 (.079)	.009 (.016)	.006 (.011)	-.009 (-.016)
Home Ownership	.065 (.047)	.036 (.026)	.041 (.047)	.024 (.027)	.019 (.021)	.170* (.092)	.096 (.052)	.084 (.045)
Number of Years at Present Residence	-.001 (-.015)	-.001 (-.013)	-.003 (-.083)	-.003 (-.075)	-.003 (-.074)	.017** (.237)	.019** (.255)	.019** (.267)
Number of Children at Home	.068** (.152)	.070** (.158)	-.008 (-.030)	-.008 (-.028)	-.019 (-.066)	.046 (.078)	.048* (.081)	.038 (.065)
Proportion of Blacks in Neighborhood		-.002 (-.095)		-.002* (-.163)	-.002 (-.130)		-.009** (-.303)	-.008** (-.262)

Neighborhood Job Prestige		.005 (.037)		-.002 (.019)	-.002 (-.026)	-.001 (-.007)	-.002 (-.008)
Race × Proportion of Blacks in Neighborhood				.003* (.263)	.003 (.221)	.015** (.567)	.014** (.507)
Membership in Neighborhood Associations					.154** (.244)		.171** (.129)
Use of Neighborhood Organizations							.301** (.143)
Constant	-.129	-.227	.351	.160	.186	.998	979
R^2	.060	.069	.040	.052	.107	.133	.176

*p < .05, one-tailed test.
**p < .01, one-tailed test.

and fear of crime have been noted as prime sources of discontent with neighborhoods (Thomas and Hyman, 1977) respondents were asked first if they or members of their families had actually been victims of crimes in their neighborhood and second whether or not they saw their neighborhood as a safe place to live. Finally, on a more positive note, we asked whether or not individuals saw their neighbors as helpful.

Looking at the first equation, we can observe that the proportion of the total variance explained by the individual level variables taken as a whole is quite small.

The most important single effect on perception of neighborhood quality, however, is the interaction between individual race and racial composition of the neighborhood. For whites, but not for blacks, negative evaluations of neighborhood quality increase substantially as the proportion of blacks in the neighborhood is increased. Running this equation separately for blacks and whites, the unstandardized partial regression weight for the proportion of blacks in the neighborhood is -.038 for whites and .005 for blacks. It is also worth noting at this juncture that this relationship holds despite the fact that a measure for neighborhood job prestige is also included in the equation. These findings reinforce the observation made by Berry and Kasarda that, "for the bulk of white American society, race continues to play an important role in ascribing group as well as neighborhood status, even after education, income, and occupational levels are considered" (1977: 220). It is also interesting to observe, however, that although the interaction between individual race and racial composition of the neighborhood had effect on two kinds of neighborhood involvement—use of neighborhood organizations to get fair treatment and number of persons known in the neighborhood—it does not have any indirect impact on overall assessment of neighborhood quality through these variables. The introduction of all of the neighborhood involvement variables does not reduce the size of the racial composition coefficient (equation 3), nor do these various kinds of involvement have a direct impact on perceptions of neighborhood quality by themselves. This suggests, therefore, that the hope of many reformers that neighborhood upgrading can come about through increased participation may be overly optimistic. Rather, our data indicates that ecological and demographic facts in themselves, especially racial composition, have a much more important effect on the way that citizens see their neighborhoods.

Equation 6 shows that neighborhood conditions, or at least citizens' perceptions of these conditions, have a substantial impact on perception of neighborhood quality over and above the ecological and

Table 13.2
Regression of Perception of Neighborhood Quality on Individual and Neighborhood Characteristics, Neighborhood Conditions, and Involvement in the Neighborhood (N = 449) (Unstandardized Coefficients Are on Top; Standardized Coefficients Are on the Bottom)

	EQ1	EQ2	EQ3	EQ4
Household Job Prestige	.009 (.062)	.005 (.036)	.005 (.034)	.002 (.015)
Race	.260 (.066)	-.637 (-.161)	-.578 (-.146)	-.333 (-.084)
Education	-.082 (-.064)	-.155** (-.120)	-.154** (-.120)	-.090* (-.070)
Home Ownership	.738** (.173)	.316 (.074)	.304 (.071)	.219 (.051)
Number of Years at Present Residence	.012 (.072)	.017* (.101)	.014* (.086)	.008 (.045)
Number of Children at Home	-.073 (-.054)	-.048 (-.035)	-.054 (-.040)	-.021 (-.015)
Proportion of Blacks in Neighborhood		-.036** (-.537)	-.035** (-.522)	-.024** (-.356)
Neighborhood Job Prestige		.059* (.131)	.060* (.132)	.040* (.089)
Race × Proportion of Blacks in Neighborhood		.040** (.629)	.038** (.599)	.024** (.380)
Membership in Neighborhood Associations			-.014 (-.005)	-.101 (-.030)
Use of Neighborhood Organizations			-.049 (-.010)	.130 (.027)
Number of Persons Known in Neighborhood			.137 (.059)	.055 (.024)
Victim of a Crime				.285* (.071)
Perceived Safety of Neighborhood				1.982** (.386)
Perceived Helpfulness of Neighbors				1.183** (.225)
Constant	4.955	4.187	4.056	2.091
R^2	.054	.249	.252	.476

*$p < .05$, one-tailed test.
**$p < .01$, one-tailed test.

demographic contextual factors. The amount of variance explained by victimization, perceived safety of the neighborhood, and perceived helpfulness of neighbors is almost half of the total variance. At the same time, however, we should note that the interaction between an individual's race and the proportion of blacks in the neighborhood has a substantial indirect effect on these perceptions of neighborhood conditions. This indicates, therefore, that the strongest indirect effect of race on citizens' perception of neighborhood quality is through its effect on residents' feelings about their personal safety and the helpfulness of their neighbors (further analysis along this line is found in O'Brien et al., 1981 as well as Chapter 16 in this volume).

CONCLUSION

As noted at the outset, the declining economic importance of central cities in the northeast and north central regions, the displacement of higher status persons with lower status persons in these cities, and the negative experiences of the neighborhood organization movement during the 1960s and 1970s lead us to believe that policy makers will focus their limited resources on working- and middle-class neighborhoods that have a high probability of success in revitalization efforts. Nevertheless, we have proposed that even with this limited focus policy makers will still have to make choices about where to allocate scarce resources and that this will tend to place more emphasis on identifying nonmaterial dimensions of neighborhood viability.

Results from the Akron Neighborhood Survey, however, point to the discouraging conclusion that at least for white residents neighborhood viability is likely to continue to be tied to the racial composition of the local area. The crux of the matter perhaps may be best conceptualized in terms of Downs' description of the process by which the value of a neighborhood is affected by the "inescapable interdependency" of its residents (1981: 16-17).

It appears that when black and white residents are in racially mixed neighborhoods, white residents do not trust their black neighbors to keep up their part of the bargain with respect to maintaining whatever is necessary to preserve the quality of their neighborhoods.

The aforementioned raises a serious ethical question and public policy dilemma insofar as racial homogeneity of the neighborhood may very well enhance its viability and long-term stability, but at the same time reinforce traditional urban patterns of racial segregation with all of their associated human costs. Recently, these issues became the center of a rather stormy debate among members of the National Commission on Neighborhoods. Several black leaders, including

Benjamin Hooks of the NAACP and Maynard Jackson (mayor of Atlanta), took issue with the de facto racial segregation implications of the presumed relationship between racial or cultural homogeneity and neighborhood viability. They, along with others, suggested that more efforts should be devoted to finding ways to build greater bonds of solidarity between blacks and whites in racially mixed neighborhoods, especially in terms of focusing on residents' common interests vis-a-vis class and associated economic issues (for a discussion of this debate see Barry, 1980).

Finally, to close on a somewhat positive although at the same time paradoxical note, neighborhoods may actually receive some real benefits from the current cut backs in public services by the Reagan administration. We noted earlier that one of the most serious obstacles facing neighborhood organizers during the 1960s and 1970s was the fact that they could not effectively compete with public service bureaucracies in providing services that might serve as inducements to get persons more involved in their neighborhoods. Although it is unrealistic, and from our point of view undesirable, to expect a return to the type of social service delivery systems extant during the immigrant-organizing era of the nineteenth and early twentieth centuries, it does seem possible that wel will see selective transfers, if only by default, of service provision from public bureaucracies to indigenous neighborhood organizations. This, in turn, would create incentives for individuals to become more involved in their local neighborhoods. (For some suggestions of strategies for transferring specific services from public bureaucracies to neighborhoods, see Downs, 1981: 179-182; O'Brien, 1975: 202-222; Ostrom, 1974).

REFERENCES

BARRY, J. T. (1980) "The national commission on neighborhood: the politics of urban revitalizations," pp. 165-187 in D.B. Rosenthal (ed.) Urban Revitalization. Beverly Hills, CA: Sage.

BARRY, B.J.L. and J. D. KASARDA (1977) Contemporary Urban Ecology. New York: Macmillan.

——— (1976) "The counterurbanization process: urban America since 1970," pp. 17-30 in B.J.L. Berry (ed.) Urbanization and Counterurbanization. Beverly Hills, CA: Sage.

CASSIDY, R. (1980) Livable Cities: A Grass-Roots Guide to Rebuilding Urban America. New York: Holt, Rinehart & Winston.

DOWNS, A. (1981) Neighborhoods and Urban Development. Washington, DC: The Brookings Institute.

FISCHER, C., R. M. JACKSON, C. A. STUEVE, K. GERSON, L. M. JONES and M. BALDASSARE (1977) Networks and Places: Social Relations in the Urban Setting. New York: Free Press.

FREY, W. H. (1979) "Central city white flight." American Sociological Review 44 (June): 425-448

GOETZE, R. (1981) "Citizen expectations can determine the success of local revitalization programs." Journal of Housing 38 (April): 205-212

——— (1979) Understanding Neighborhood Change: The Role of Expectations in Urban Revitalization. Cambridge, MA: Ballinger.

——— and K. W. COLTON (1980) "The dynamics of neighborhoods: a fresh approach to understanding housing and neighborhood change." American Institute of Planners Journal (April).

HAUSKNECT, M. (1962) The Joiners. New York: Beminster.

JANOWITZ, M. (1967) The Community Press in an Urban Setting. Chicago: University of Chicago Press.

KAPLAN, M., S. GANS, and H.KAHN (1970) The Model Citits Program. New York: Praeger.

LANGE, J. K. and D. J. O'BRIEN (1978) "Needs and formulas: operationalizing justice in community development funding." Sociological Focus 11 (October): 317-327.

LIN, N. (1976) Foundations of Social Research. New York: McGraw-Hill.

LINEBERRY, R. L. and I. SHARKANSKY (1978) Urban Politics and Public Policy. New York: Harper & Row.

MUELLER, S. A., D. J. O'BRIEN, and N. GILDER (1981) "Racial composition and informal social networks: a contextual analysis." Presented at the annual meeting of the North Central Sociological Association. April 23-25, Cleveland, OH.

NATHAN, R. P., P. R. DONNEL, S. F. LIEBSCHUTZ, D. MILTON, and associates (1977) Block Grants for Community Development. Department of Housing and Urban Development. Washington, DC: Government Printing Office.

O'BRIEN, D. J. (1975) Neighborhood Organization and Interest-Group Processes. Princeton, NJ: Princeton University Press.

———, J. K. LANG, and A. KEEVER (1981) "The effects of racial composition on residents' evaluations of their neighborhoods." Presented at the annual meeting of the American Sociological Association, Toronto, August 25-28.

OLSEN, M. E. (1970) "The social and political participation of Blacks." American Sociological Review 35 (August): 682-697.

OSTROM, V. (1974) The Intellectual Crisis in American Public Administration, University Station: University of Alabama Press.

SANDERS, H. T. (1980) "Urban renewal and the revitalized city: Rosenthal (ed.) Urban Revitalization. Beverly Hills, CA: Sage.

SCHOENBERG, S. P. and P. L. Rosenbaum, (1980) Neighborhoods that Work: Sources for Viability in the Inner City. New Brunswick, NJ: Rutgers University Press.

THOMAS, C. W. and J. M. HYMAN (1977) "Perceptions of crime, fear of victimization, and public perceptions of police performance." Journal of Police Science and Administration 5 (September): 305-317.

U.S. Bureau of the Census (1981) 1980 Census of Population and Housing, Advanced Reports, Ohio. Washington, DC: Government Printing Office.

U.S. Department of Transportation (1977) An Introduction to Travel Demand Forcasting: A Self Instructional Text. Washington, DC: Government Printing Office.

VERBA, S. and N. H. NIE (1972) Participation in America: Political Democracy and Social Equality. New York: Harper & Row.

WARREN, D. (1977) "The functional diversity of urban neighborhoods." Urban Affairs Quarterly 13 (December): 150-180.

14

Enduring Elements of Urban Design and Architecture

RICHARD FLEISCHMAN

☐ OF ALL THE ISSUES THAT CITIES CONFRONT as they grow and mature, urban design is probably the most challenging and the most enduring. Urban design, the process of weaving together a diverse array of structures into a distinctive city fabric, is a very highly sophisticated art form. Cities are statements of civilizations and the quality of urban design reflects a city's success at forging community values and agreements on how development is guided or shaped. The principles or aesthetics that guide urban design are what allows some cities to achieve greatness.

One building standing alone in an open landscape is experienced as a work of architecture. This free-standing building, regardless of silhouette or surface architecture, is an object in space, and seldom contributes to urban design. However, organize buildings together, and another art form is possible. Several things begin to happen in the composition; they create what could not be possible for the isolated building. You may have the opportunity to walk through and past the collection of structures; as the path changes direction, new and unusual experiences of an entirely different mass, profile, or facade can occur as a result of an unsuspected or unknown location. Beauty in architecture and cities is largely dependent upon their harmonious and artful relationship to space. A good example of this is shown in Exhibits 1 and 2, which represent Terminal Tower and Public Square in Cleveland.

To design urban space as a forceful display of architectural creativity and design it so that it can be used and occupied by everyone

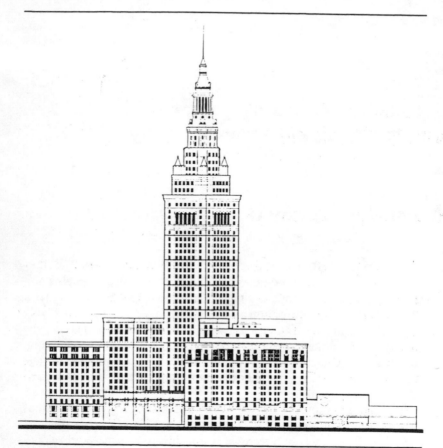

EXHIBIT 14.1

in a relatively unselfconscious manner requires a sense of balance not unlike Schopenhauer's description of the beauty that exists in architecture—it occurs only when gravity and support are equal forces.

One can observe the history of a city as a scaffold to which ideologies reflected as architecture and urban design are attached as a dynamic reservoir from which alternative ideological interests emerge or are renewed. It is clear, from the initial development of cities, that people have added, remodeled, and subtracted from their community setting in the most dynamic manner, and the style of architecture had little to do with its success. The architecture of the city is a diverse and often complex text of visual symbols. It has a universal relationship to the city, it is unique because it is unlike other art forms, it is

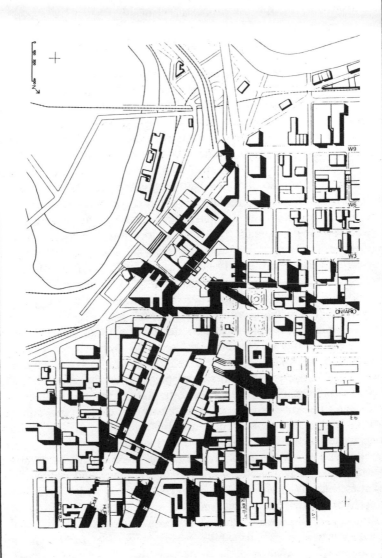

EXHIBIT 14.2

251

scrutinized in full view, it can neither control nor escape the city's form and boundaries. It can be identified as an art of intervention and the creative relationships that result are integral to its success.

PRINCIPLES OF URBAN DESIGN

The complexity of conformance and unity increase with the size of a city. However, the principle of urban design remains. Basically, it involves an informative and visually artful presentation of the city and its architecture, the whole time maintaining a sense of orientation for the user. Orientation is simply a matter of place, knowing where you have been, where you are, and where you are going. As the size of the city increases this has become more difficult. Landmarks become more important and so should be given greater considation both in their location and in the design of the spatial corridor from which they are seen. For centuries the basilica, the campanile, the cathedral, or the castle was the influencing edifice; its location was critical to the composition. In the modern idiom the focus has changed to the office tower or highrise complex. The strength of the corridor of space becomes critical when it in itself becomes a sense of place. This sense of orientation is not provided when the containers that define the space lack quality by being totally anonymous.

But a primary principle of urban design is a focus on the space created by buildings as equal to the buildings themselves. A city is generally thought of in terms of size, its population, and physical extent. Size is closely linked to shape, physical outline in horizontal plan form, and vertical profile or contour. Size, shape, and surface are further modified by density. The intensity of use of the land is measured by the number of people and buildings. The bustling urban centers are magnets of the city: People are the generators.

Good urban design must inspire the use of space and attract people. Unquestionably it is the concentration of buildings and people that give cities their dynamic quality. Suburban sprawl which invariably surrounds the center contradicts recent planning scenarios that search for a more urban consolidation. European cities have traditionally followed a pattern of density while dispersing by necessity. Expansion does not mean to disregard the scale and direction of what exists. (Scale being the difference of what is and what appears to be.)

Urban design may be regarded as a synthetic inventure mapping of physical conditions establishing and exploring areas of the city. It is the further definition of the architecture of space. It is the critical combination of general notions of urbanism and specific concepts. The intent of urban design is to seek a logical and strategic articulation of elements inherent in an exciting condition. Unique to this spatial organization is the invention of walls, streets, public spaces, and building typologies that resolve existing ambiguities in the form of the city. Exhibit 3 represents such a treatment of downtown Cleveland.

Urban spaces, like architectural spaces, therefore may be self-contained islands unrelated to other spaces, or may be interconnected and best appreciated by moving from one space to another. The medieval cities provide excellent examples of pedestrian walkways culminating in the square or piazza dominated by the cathedral. Renaissance cities are described by a classical concept relating more to the geometrical symmetry of squares and boulevards than to the informal, almost spontaneous pedestrian character fo the medieval towns. The twentieth-century city is affected by our technology, the automobile, and the corridor that permits this mechanized movement from center to center. New perceptions of mobility must be constantly discovered.

The downtown parking garage, for instance, which Jim Rouse has indicated has been essential to the revival of American downtowns, must be reconceptualized as a symbol of this new mobility. These garages have been hidden, ugly, and poorly conceived. It is time to refocus upon the value that can be contributed to the use of down town spaces by well-conceived, well-placed, and safe parking concentrations. Since many Americans worship the automobile for the convenience, mobility, privacy, and even intimacy provided by it, perhaps we should treat its midday resting place with more respect.

Transportation planning will be one of the most critical issues of the next two decades. Regardless, society must continue to respect the pedestrian and the need for human scale. Both technological and humanistic expectations can be realized by the skillful design of buildings and their massing. Understanding the expanding technological and societal changes is essential to a visionary attitude, and can contribute to the success of urban spaces. It is fundamental to the design process that the size, shape, and surface treatment of the container or the enclosing buildings be defined in the urban space. Proportion of the spatial definition is essential to creating urban design. Groups of buildings, clusters, ensembles whose facades are sensitively juxtaposed can convey a sense of place, an excitement, and

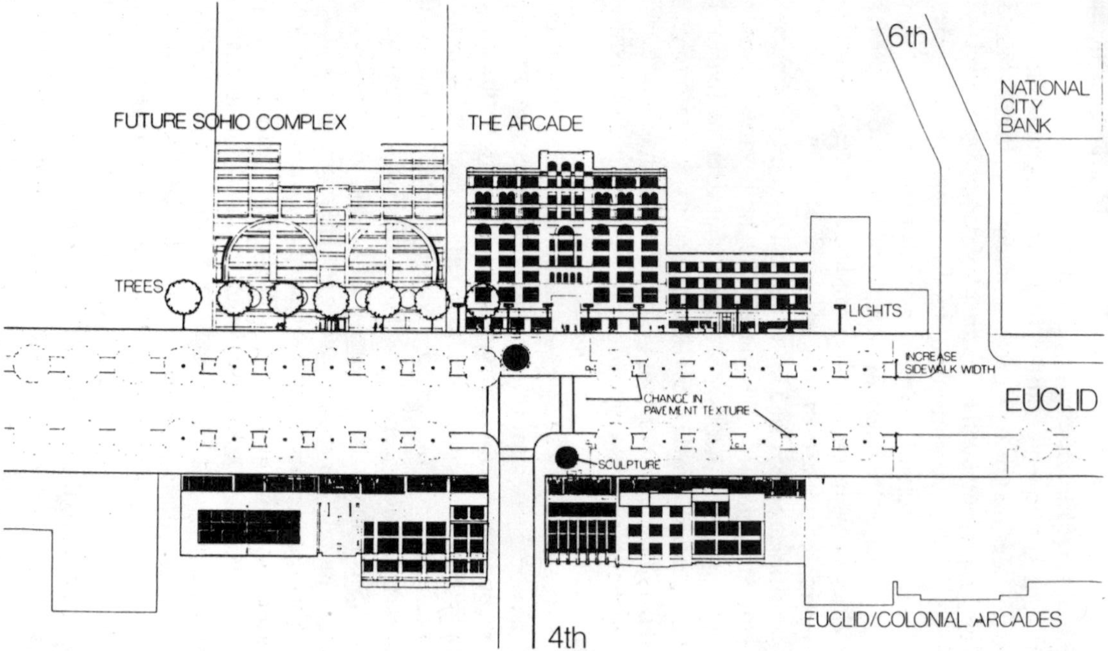

EXHIBIT 14.3   Street Guidelines/Detail

254

vitality that are ingredients that determine success. Exhibit 4 provides a perspective of the skyline of Manhatten that reflects three centuries of successful urban design.

Cities are where you find the energy, the vibrance, the vitality that denotes the human community. Cities have always had this meaning for people. An excellent example is Rome. This city acted as a forum where people communicated with one another, where art and culture, government, commerce, industry, and religion were born and were sustained through states of social and technological progress. The city in general is a space for opportunity and agreement. It is dedicated by each of us in exchange for community. This innovative exchange provides a sense of rapport, of commonness, all components act together and are a natural celebration. When this happens, it can be an inspiration with the promise of the possible. The creative process and the techniques for urban design and their invention are similar to those associated with architectural design; the difference is the scope of investigation requires a broader understanding of societal needs while maintaining sensitivity to form and detail. This is the ultimate principle of urban design.

CONTEXTUALISM

Contextualism provides the framework for the design approach applied in urban conditions. This attitude takes into account the existing qualities of streets, topology, fabric, and facade. Understanding the concept of the context is fundamental to coherent design and consistent strategies. The grid pattern was prominent in American and European cities of the late eighteenth and nineteenth centuries when abstraction exhibited one or two basic attitudes. In fact, the ground could be seen as a plane with a two-dimensional street grid superimposed and buildings as objects arranged upon the gridiron plane. The second, and inverse, attitude assumes the city to be a three-dimensional grid of buildings with streets carved out of it. As Exhibit 5 illustrates, we need to look at urban form as a piece of complicated sculpture.

Urban design requires an expanded effort, a different process from designing a building— a process involving government, community, investors, and entrepreneurs. The sequence of events is more complex than the creation of a single structure. Consider a planning philosophy for urban consolidation, an approach to urban development in opposition to the currently progressive unicentered city with its high density, highrise commercial downtown core, and its sprawling

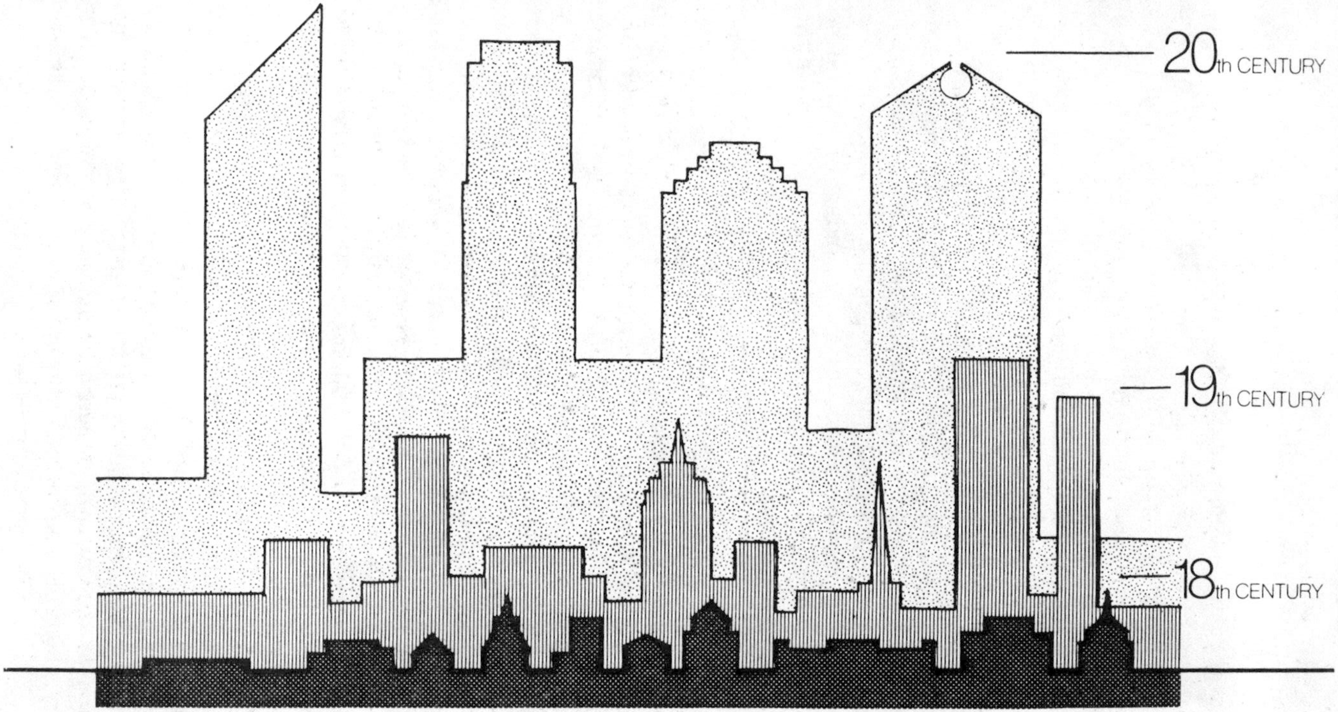

20th CENTURY

19th CENTURY

18th CENTURY

EXHIBIT 14.4

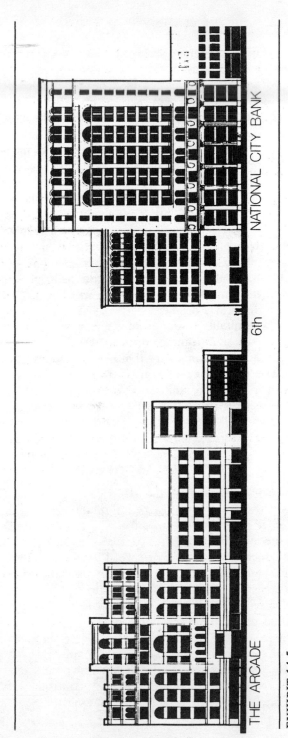

THE ARCADE   6th   NATIONAL CITY BANK

EXHIBIT 14.5

suburban periphery. This alternative advocates conserving and building on the existing urban fabric and addresses the theme of city segments. Mixed-use buildings would be encouraged. Multiscale, multisize projects would be encouraged as infills marking the existing fabric. Each intervention attempts to reinforce the idea of urban consolidation through more even distribution of densities and respect for particular context, and to propose an attitude that might return to our cities the architectural continuity and urbanity they once had.

Facade articulation will create heroic or human scale, depending on the size and degree of texture intrinsic with the character of the building. The various elements that are a part of the composition for St. Peter's Basilica are heroic in size and therefore convey a monumental effect. The design for the Petit Tremont portrays an entirely different scale. The difference is striking, the size variation inconceivable, particularly when you compare the height of St. Peter's (Exhibit 6) to the United Nations Building in Manhattan. Yes! Both structures are identical in height!

The contextual quality of our cities is a relatively new element that has emerged as a significant component that describes traditional detail and character. It is apparent that new developments with newer techniques and technology of construction can readily create new options with new opportunities. Professionals who employ architectural principles as well as urban design criteria have initiated innovative planning concepts that produce infill solutions while maintaining the historical character of the city.

DOWNTOWN AS CATHEDRAL

Downtown is the center of the American City. Louis Kahn has written that "the center is the cathedral of the city—to embody this role, the center must also be the acropolis of the city." Downtowns develop at the most strategic place of access in our cities, and contain the highest concentration of services for the whole community. Even though this function begins to diminish and the virtues of our downtown lessen as communications improve, there appears to be a resurgence to retain the valuable facilities of our downtown. It must be understood that change and development is the key to sustaining and revising the central business district. Sound economics and ample and direct access are elements within the planning cycle that help reinforce this concept. Most American downtowns are coming back.

Buildings are costly and less flexible or controllable than open space, and should be seen first as the containers, liners, and fabric

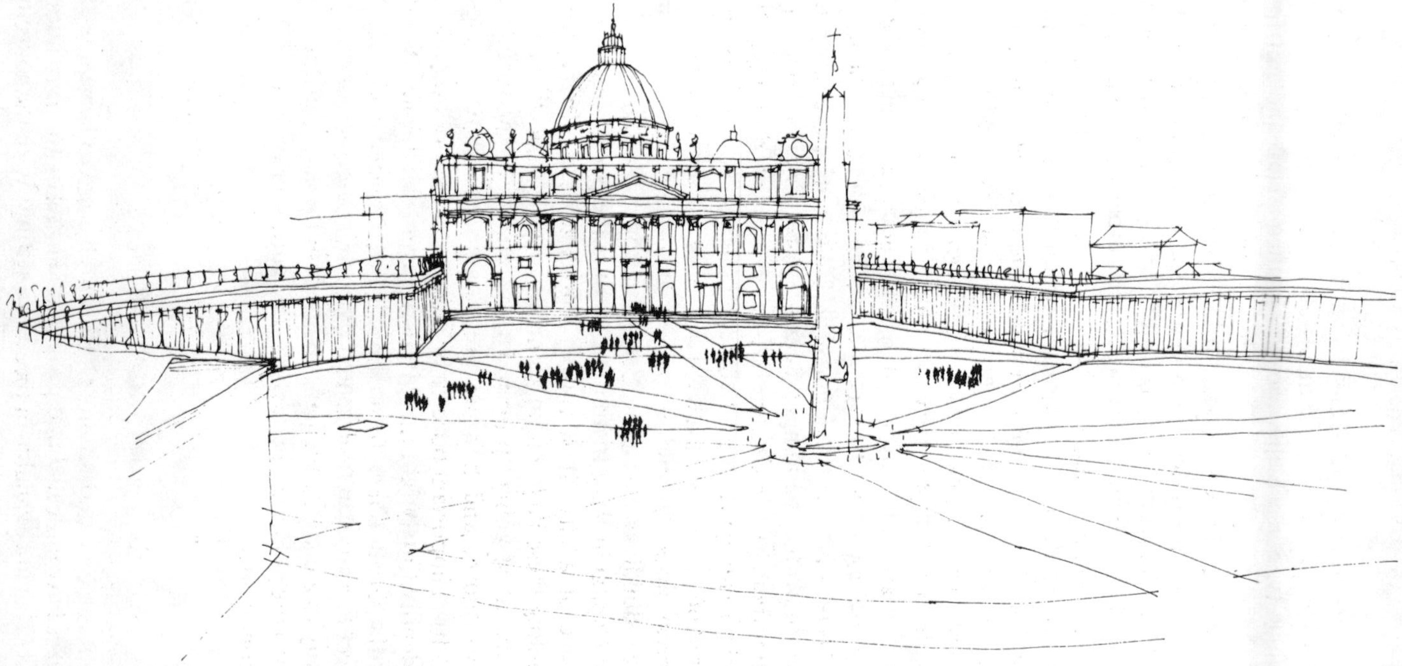

EXHIBIT 14.6

259

attached to the cities' scaffold. Most buildings in cities are not intended to be special, they are most successful as repetitious variation on a single type in which variety is in their detail and not in their overall shape. Specific buildings should occupy special locations and serve special purposes. When all, or too many buildings are special, or unique, the result can be urban chaos and confusion. It is apparent that bad buildings, conceived as objects in themselves with little or no regard for their surrounding context, their urban design value, or their important function, act as bad performers. This situation is exaggerated by the urban renewal movement where the existing and surrounding city was considered contemptible, out of date, even corrupt. Buildings were to be ignored, or even destroyed.

Needless to say the best cities are characterized by homogeneity of type, scale, detail, and variety of points of function. Changes in this architecture are most successful when gradual and incremental attention is being paid to the practice, concentrations, and styles of the place. *There is an art of relationship.* Objects in space relate to one another. Individual buildings, regardless of their appearance, play a very important role in the total visual expression of a city. Objects with unique silhouettes are special components in the organization of the numerous parts. A spectacular profile can be visually effective from great distances and identified as terminations of a spatial sequence. Buildings such as St. Peter's in Rome, St. Mark's in Venice, and the Capital in Washington D.C. affect the perception and visual experience of a city. It is further enriched by a variety of views surrounding these large major elements in their respective cities. We are cognizant of going up or down, the quality of the surface upon which we move, and how we are transported. Natural landscape features form important borders or edges of cities. For example, three rivers identify Pittsburgh, the Patapsco River basin locates Baltimore, two of the Great Lakes form the edge of Chicago and Cleveland. Entrances to a city can be accented by portals, the doorways to a city, a gateway to a district. We have focused on the center of the city. At best this core contains the highest concentration of services for the whole community.

AMERICAN MODELS

Recent success in Baltimore, Pitsburgh, and other large urban complexes have indicated that people need to relate to the richness and resources that are available in the urban setting. With the collection of new structures in the cities mentioned above, the streets and squares perhaps more than any other elements of urban design have had two powerful effects—they unite central spatial enclaves together and support the function of the activities that line the spatial corridor.

Consideration of the beauty of the cities of the past is essential, for it provides encouragement as we envision the future. We must be cognizant of, and sensitive to, the methods associated with the numerous accomplishments of the groups or individuals responsible for the successes. It is logical to begin by adapting the philosophy that to create a beautiful city, we must start with a beautiful concept and the conviction that every city will have the opportunity to become beautiful. This identity does not have to be limited to a few internationally known urban centers (Rome, Vienna, San Francisco). The image can be acquired by any city in any country. The primary requirement is simple—you have to want the image. There are examples where cities that have had reputations for ugliness have initiated planning strategies that reflect and recognize the outstanding and natural beauty of their location as the design basis of its urban form.

The topography and the rivers of Pittsburgh are the natural elements that assign a unique image to the city form. However, the industrial pollution, that once blanketed the city led to urban decay. As a result of media intervention progressive urban design goals to improve the environment were initiated and carefully implemented. Drastically changed, the city now presents an entirely different appearance and life. The development of the Pittsburgh Plate Glass (PPG) building and other significant projects have helped to alter the silhouette of the city while expanding its expectations. Fortunately, the creative initiative in the design of the PPG profile is also reflected in the resultant urban space created at street level.

One of the most important initiatives in the redevelopment of Pittsubrgh was the decision by PPG to build a new headquarters. PPG was positive about the size and location of the site and required the city to use urban renewal power to assemble its land, with the city providing conditional approval. They wanted those involved to know that their innovative partnership was motivated by their interest in a new type of urban renewal and was not party to a program of total clearance. The city designated a 26-acre area that included a 10-acre historic district. Therefore, redevelopment also called for renewal and preservation of buildings. This appraoch to urban revival has been repeated in Toledo, Cincinnati, San Francisco, and Milwaukee. But Baltimore is the best example.

Baltimore stands at the mouth of the Patapsco River, a narrow fringe of the Chesapeake Bay that appears to be pushed 200 miles into Maryland. It is the largest stretch of natural inland harbor in the country. The fruits of urban planning are now being realized. Since the mid-1950s cumbersome freeway plans and other individual programs have been deflected in order for responsible redevelopers to proceed with a plan whose basic intent was to return the shoreline to the people. In 1977 the 28-story World Trade Center was the first public presence on the inner harbor. First, of course, was the waterside promenade and a series of self-effacing office towers that act as a buffer between the Charles Center and the harbor. Harbor Place opened in 1980 after the completion of the National Aquarium and ceremonial landing. It is a twin two-story pavilion containing 250,000 square feet of shops, restaurants, stores, and other assorted delights. Like other developments throughout the country, it is the inclusion of the celebrated marketplace that has brought the enthusiastic community to share in its urban opportunities. It has become glorious for people to congregate, similar to the piazza of St. Mark's. It gives a sense of place and is programmed for people.

URBAN COMPOSITION

Architects are accustomed to regard their design as large compositions that have their beginning, end, and central theme. A building design is a complex entity comparable to a piece of sculpture. In urban design, as mentioned earlier, an individual architectural project may not be relevant. These building elements must act as part of a larger composition. They are a continuous part of a pattern, they participate in a weaving process creating a fusion of entities. An urban design concept is an idea for structuring spaces, masses, and activities in a city. The architecture, regardless of its vintage, acts as a container, providing the sizes and shapes of the voids. These measured voids, or corridors of space, are created with an intent of maintaining visual connectors between, around, or through the containers—the architecture.

Rome is a city of streets and squares lined with buildings, not a city of buildings connected by streets. The architectural complex of St. Peter's is an excellent example of sequential spaces created in stages. Michelangelo created the major design for the basilica where the dome was the central element—the composition was completely symmetrical. However, Maderna's facade profoundly changed this symmetry and the visual experience, relegating the previous architectural expression as to the background of the new facade. Bernini's design for the portico diminished the importance of the facade. The elliptical portico created two cross-perspective views that put Maderna's design in parenthesis, enclosing it in another space which has its own value and thereby permitting attention to focus, once again, on the dome. It is in this context that we begin to understand the significance of the components, and when artistically orchestrated, to appreciate the aesthetics of a city—a moment of agreement, of unity, and a sense of place. But remember, even Rome began without urban design or schools of architecture and urban studies.

Venice also is a city of squares. However, instead of streets, its urban character is identified by canals and pedestrian walks. The architectural facades define the variety of linear spaces that provide an articulate network of cavities culminating in the piazza, which appears to be a stage for St. Mark's and the adjoining buildings as shown in Exhibit 7.

After observing a multitude of planning configurations, it's not difficult to recognize that it is the great visual variety, the combination of the old and the new, the accidental excitement that gives cities their character and differentiates one city from another. It is apparent that society is again responding to and demanding visionary planning in association with qualitative concepts. This, along with a combination of political and social forces should necessitate immediate action, recognizing that cities must be developed with committee leadership. As a result, we can expect urban designers recommending a reversal of the traditional planning process in order to accomplish a more humane environment. Essentially, it is a consideration of existing assets, including the character of space and structures, as a basis for future development instead of a hindrance to be worked around or ignored. Society will be drawn to quality, will support growth, and will be less apt to object to change if what is familiar remains. This actually is an extremely civilized idea. Architecturally, the way seems open to a rich mix of buildings and spaces from various periods that would give a tangible sense of city history, quality excitement, individual character, and comfortable continuity. As new structures are contemplated, they should be treated as a partner in the whole, rather than as an intruder. Tradition, which is often regarded as a cosmetic, can be a basic value.

It is not a time for eclecticism, but an opportunity for concerned and sensitive design. Building complexes such as the Citicorp in Manhattan, as well as the American Telephone and Telegraph building, present this thought process and are excellent examples of

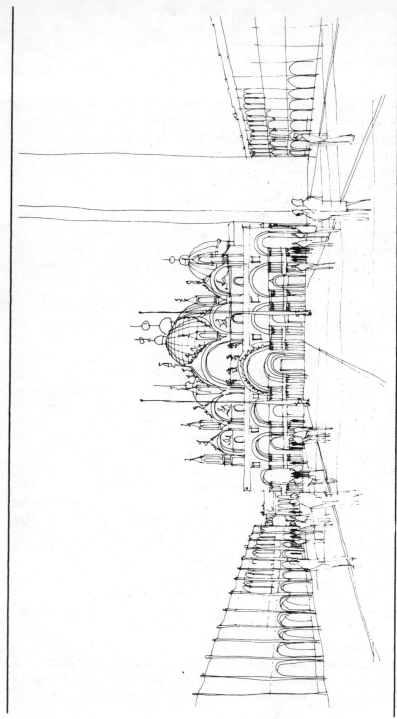

EXHIBIT 14.7

264

the proposed partnerships between business and urban design. Both are unique, but appear to provide compatibility even with their distinctive character. Their silhouette also characterizes the direction new building design will pursue as redevelopment continues in the central business district. As we contemplate planning for 2001, we become more aware that tradition requires understanding. Its influence can initiate new opportunities for createive design. Buildings of different periods will maintain a value and even combine with redevelopment to maintain economic interest.

Our mobile society will not diminish the contextual conditions resulting from this composition; it will reinforce the visual form of the city. The pedestrian is the real city client. Vehicular traffic and accessible parking are important considerations in recommending urban design guidelines for the city. The concentration of people, regardless of the individual objectives (worker, shopper, visitor), is what urban design seeks to achieve. Pedestrianization is a major goal of urban designers. However, this does not mean vehicle-free cities.

Urban design is undoubtedly an ambitious undertaking, in some regards the most critical of the design arts. It is an enormous undertaking to design large areas of a city. Who can predict what cities should be? It is clear that an enduring element in urban design during the next twenty years is maintaining the contextual character that permeates downtowns. Fiscal restraints, economic uncertainties, and formidable existing buildings are all parts of the recycling theme that prevails in all parts of the country.

Let's consider both Washington, D.C. and Cleveland. Washington is a lively metropolis. Its image as a hub for governmental activity has continuity, and has not diminished. This has occurred without changing L'Enfant's basic plan. Critics often attacked the wide distances that seem to relate only to moving traffic and military parades. The grandiose plan incorporated a vision and a solution that translated the congestion that occurred in other urban areas. The predominance of the Capitol is guaranteed by the Congress's 1910 Height Limitation Act. Therefore, Washington is a radiant city that is livable and human. Even though the enthusiastic architect is refrained, the street plan and height limit definitely determine the city's identity. As an expanded sense of place, Washington D. C. definitely displays that character. The Capitol or the Washington Monument are always visually available.

L'Enfant designed Washington for 800,000 people living in four-story houses. The city now has a population of 600,000 and the zoning envelope is limited to ten stories. The urban design guidelines, initially appealed by critics, are now translated into quality and freedom. Even though there are office towers, Washington has its own description of contextulization. It has combined the dignity of Paris with the ego of aggressive democracy.

Cleveland is enjoying a similar community plan agreement with the Standard Oil Company. This major firm chose to locate its international headquarters very close to where the firm was originally founded by J. D. Rockefeller in the center of the city. Cleveland designated the Euclid/Prospect Action area to be investigated for blight and commissioned consultants to create a community development plan (CDP). This is a new approach. The city council, after reviewing the survey material, passed legislation that permitted the acquisition of the land by "eminent domain" but also encouraged the property and building owners to personally participate in the CDP for their own benefit. The final draft describes the options that are available to developers and others who want to participate in the redevelopment of the retail core area of Cleveland. The plan includes the extent of retail that should be maintained, location and height of office towers, hotels, and a critical use of parking corridors as mid-block that will facilitate the use of the area. Guidelines are now being developed that will coordinate these efforts and will graphically describe the contexture of these critical areas downtown. This is illustrated by Exhibit 8.

CONCLUSIONS

There are now disasters occurring in major cities around the world. Paris and London are excellent examples of the exploitation and disregard for the vintage and existing images. But the urban renewal and removal disasters of these cities that occured within the last twenty years have inspired new, strict standards.

It is essential that we create zoning envelopes and enact urban guidelines that will preserve the historic continuity, charm, and livability of our cities. Wollf Van Eckardt said, "After all, thousands of people can work thousands of years to make a city beautiful. But it takes one greedy developer only one year and one large skyscraper to spoil it all."

Figure 1 and Exhibit 9 are useful illustrations to end this chapter. Exhibit 9 provides an utopian perspective of a multileveled urban concentration that respects the different scales of urban space. Grand Central Station and Rockefeller Center in New York still remain the exemplars of this kind of urban design impulse. (Terminal Tower in

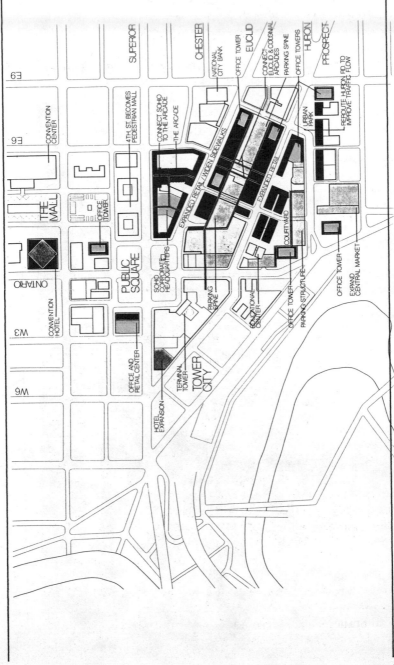

EXHIBIT 14.8

267

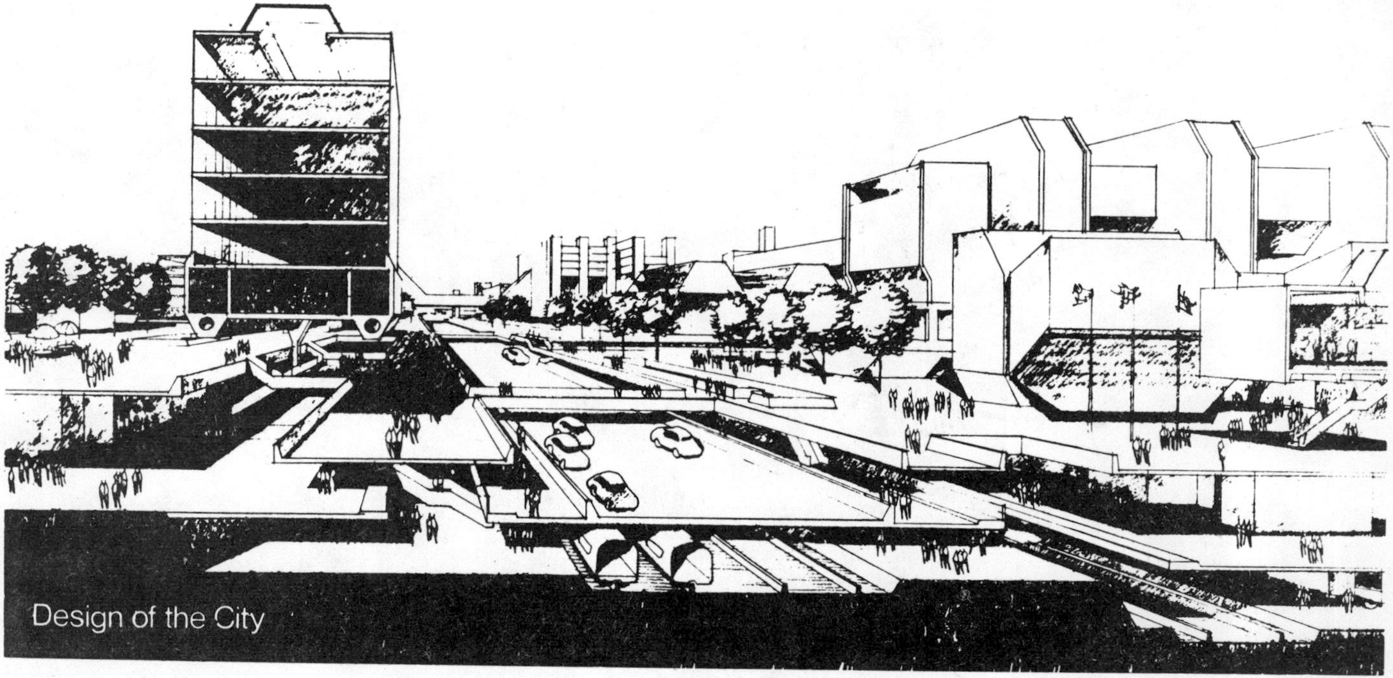

Design of the City

EXHIBIT 14.9

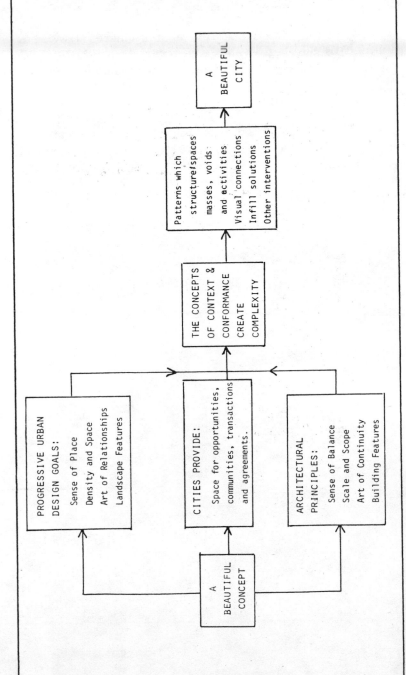

FIGURE 14.1

Cleveland was a precursor, still incomplete, of the Rockefeller Center form.)

Figure 14.1 attempts to synthesize the issues elaborated in this chapter. Beautiful cities must begin with beautiful concepts, and must follow the primary principles of both urban architecture and urban design. Together they create a concept of the context in which usable patterns of urban space are realized.

The challenge of the next decade or so is to create new kinds of interventions, particularly infill designs, that will contribute new progressive forms attached to the prevailing fabrics of our evolving urban centers. This process will succeed if we stop just building buildings and start designing cities.

15

Frontierism and Civic Leadership

FRANK COSTA
YONG CHO

☐ CONCEPTS OF TIME and human conditions vary from place to place. In many Eastern cultures time is viewed as a cycle that returns back to a point of beginning. Recurring events and periods, such as the seasons or birth and death, lend strength to the cyclical view of time. In the Western world, on the other hand, time is viewed as linear movement. The passage of time is absolute. Time lost is never regained. This Western view has been coupled with another peculiarly Western concept, the concept of progress.[1] Progress develops as time unfolds. These twin concepts of time's passage and progress are so intrinsic to us that we seldom concern ourselves with them until they are challenged (Banfield, 1974).

Progress is tied to the concept of seizing opportunities when they arise. Opportunities in American life have, historically, been associated with the frontier experience. The establishment, growth, and apparently inevitable decline of cities in the United States is also tied to the forces created by frontierism.

It is the premise of this chapter that there are four interrelated factors or conditions that affect or stimulate change in American urban settings. These are:

(1)   The nature of civic leadership—What are the social and economic orientations and goals of the dominant leadership groups in the community? Privatism and its variant, paternalism, are the dominant leadership types. They espouse values and behavior patterns favoring private initiative for personal or private gain, often at the expense of the community.

(2)   Civic morale—A condition that reflects citizen attitudes toward social and economic opportunities available to them and the equity by which these opportunities are apportioned.

(3)   Urban vitality—A community with good or adequate morale can be termed "vital." People perceive that social and economic opportunities exist for them.

(4)   Frontierism—Seeking opportunity in a new geographic setting.

Each will be discussed in turn.

CIVIC LEADERSHIP

Civic leadership can be defined as the directive force in civic affairs. Those who exercise leadership can be considered a political or socioeconomic elite in the Pareto sense, that is, they control a disproportionate share of civic resources (Pareto, 1902). We identify three civic leadership types: (1) privatistic leadership, (2) paternalistic leadership, (3) civic-minded leadership.

Leadership in American cities is exercised through the control of political or social groups or through the control of group resources such as newspapers, financial institutions, and cultural institutions. Given the heterogeneity of American urban life, no leadership type exists to the exclusion of others. There is a dominant leadership type and there are subordinate leadership types.

Dominant leadership develops from the action of initiative groups, those who propose courses of action and impel the entire community to follow. Subordinate leadership exists in the form of intermediate groups and latent groups. Dominant leadership remains unchallenged as long as the civic condition remains stable. When conditions of instability emerge, then intermediate or latent groups may move forward to take advantage of a fluid political setting.

Privatistic leadership stresses economic gain as its primary goal. Initially, privatistic leadership tends to be fragmented or unintegrated. It tends to flourish in rapidly growing communities (true frontier settings). Privatistic leadership does not value community equity but tends to create conditions favorable for community opportunity in economic matters.

Paternalistic leadership is integrated privatistic leadership. It constitutes a power elite that seeks to maintain stable social conditions conducive to promoting continuing economic growth. Under conditions of paternalistic leadership, community equity would not be fostered, but community opportunity may exist if economic conditions are expanding.

TABLE 15.1   Leadership Characteristics of Leadership Types

Civic Leadership Type	Level of Political Integration	Potential for Urban Power Sharing	Leadership Attitudes Toward Equity	Opportunity Level
Privatistic	Low	High	Low	High
Paternalistic	High	Low	Low	Low
Civic minded	High	High	High	High

Civic-minded leadership stresses both community equity as well as community opportunity but will always place its greatest emphasis on equity. It too can flourish in a variety of urban settings from growth to decline.

CIVIC MORALE AND URBAN VITALITY

Civic morale can be said to be at its highest level when opportunities for economic mobility and social improvement are high and when the access to these opportunities is high. Such a community may be said to be vital. The condition of vitality means that citizens value equity and productivity; they localize their efforts in their own community while ensuring that there is reasonable justice among groups so that members of all groups wish to engage in activities that benefit the community. These joint activities under most circumstances will produce local benefits.

The condition of obsolescence presents factors that are the reverse of vitality where concerns for community equity and the attitudes toward community productivity begin to erode. In these circumstances none works to save his or her city, since such persons are measuring the fate of the place in which they are living only by the rule of economic comparison; and in the case of such individuals there is no reason for those who can move to stay. This is especially true of older, poorer central cities and their urban fringes with absentee owners and managers. In some cases the fringes maintain an apparent vitality for some time at the expense of the core, either by abandoning it or by cannibalizing it; however, ultimately the fringes themselves become part of a regional decline as all of the cities and all of the areas within the region suffer continued social and economic comparison to other areas.

FRONTIERISM

Turner called the frontier the major motivating force in American history (Turner, 1920). Most of the energies and innovative technologies were developed in response to the developing American frontier. Opportunities inherent in the frontier propelled the innovative surge in transportation and agriculture during the nineteenth century. Turner lamented the closing of the frontier toward the end of the nineteenth century, but American economic development continued long after. Elazar accounts for this continuing growth through the concept of recurring frontiers. He describes three frontiers in American life. These are the universally accepted rural frontier of which Turner wrote, an urban frontier, largely ignored by historians, and a technological frontier that sought to combine aspects of both the rural frontier and the urban frontier. he successor to the original rural frontier was the urban-industrial "opportunity space" of the early and mid-nineteenth century. Old cities grew and new ones were established; developing industrial systems required concentrations of land uses and people flocked to them from the countryside and overseas to work in the factories. There were opportunities for economic and social achievement in the cities and, as the cities grew, they were in fact the new frontier (Elazar, 1970).

Eventually the nineteenth-century cities as frontiers gave way to an even newer frontier when, after World War II, a "metropolitan frontier spread across the nation in a generally uniform pattern and was emphatically manifest in the rapid postwar growth of suburbs" (Elazar, 1970). The metropolitan frontier developed as a result of technological changes in transportation and the need for Americans to come to terms with their social values and their economic needs. Transportation technology greatly enlarged the radius of efficiency for an urban worker. Where once he was required to live close to work, frequently within walking distance, new developments in highway and automobile technology allowed him to live further out and still retain his economic ties to the workplace. Thus, the metropolitan frontier was created.

In the 1980s new frontiers are being created that foster new opportunities for those drawn to the frontier settings. Where formerly on the metropolitan frontier opportunities for new growth existed in the same region, today a new frontier of regional competition and conflict has been created. The areas of opportunity are to be found in the South and West, and a migration pattern that was previously intraregional has now become interregional. The perception current today is that economic opportunity is to be found in the Sunbelt and upwardly mobile people are moving out of the Northeast and Midwest and into the South and West.

Finally, an incipient transnational frontier can be seen developing. This frontier does not recognize national boundaries but localizes investment anywhere in the world where economic opportunity seems likely. One way in which the developing transnational frontier differs from previous frontier settings is in the relative inability and unwillingness of Americans to move with it. The transnational frontier seems to be the special creation of multinational firms that transcend local or even national allegiances in their search for opportunity space.

The frontier mentality has given Americans an inherent sense of optimism about their futures. If things do not seem to work in one place, then we will find a new place of opportunity. Frontierism has fueled most of the migrations of American history—onto the farm, from the farm to the city, from the city to the suburb, and now from one economic region to another. The optimism of the frontiers people provides a sharp contrast to the pessimism of those left behind.

Frontier settings are, essentially, focal points for resource exploitation and nonfrontier settings provide environments for underutilization of already created resources. In both cases waste results. The creation of costly new urban infrastructure (sewers, utilities, roads) in newly opened suburban areas goes on, while systems already in place in the central city are abandoned. Frontierism tends to foster a sense of impermanence because new settings for living are continually being sought. The old neighborhood deteriorated in the fifties and sixties; the old *region* is deteriorating in the eighties.

Throughout our history mobile Americans have been able to avail themselves of frontier opportunities. However, with the gradual development of a transnational frontier, few Americans have the resources or skills to exploit this new setting. For the first time Americans may find that they cannot move on but must remain fixed in a declining national setting. For the first time, optimism about the future may give way to pessimism. Frontierism as a political and social safety valve may no longer be a force as it has always been in diffusing political and social pressures for economic redistribution. Issues of equity, which had been muffled by opportunity, may come to the forefront of national concern.

ECONOMIC CHANGES AND
URBAN DECLINE IN
A TRANSNATIONAL FRONTIER

The decline of the traditional industrial cities of the Northeast and Midwest marks the beginning of a new era, the era signifying the end of frontierism and privatism. The recent urban decline is most clearly manifested in such indices as population loss, job loss, shrinking income and tax base, growing needs for public services of all kinds, and the fiscal constraints forcing government cutbacks.

This urban decline is a direct consequence of drastic economic changes that have created a new order of the world economy, namely, the deindustrialization of the American economy and the corresponding growth in global economic interdependency, the intensity of which is growing rapidly.

Deindustrialization does not mean that all industries suffer deterioration. Rather, it means that a wide-ranging variety of traditional manufacturing industries, including many key sectors upon which the economic might of the United States has been founded, have experienced a precipitous decline. These include steel, automobiles, tires, shipbuilding, electronics (electric) and electronic appliances, textiles, and garments among others. These industries have lost competitive advantages with foreign manufacturers in both American and transnational markets. America's productivity stagnates, prices soar, product quality falls, and products themselves lose touch with consumer demands.

Some of the more salient economic changes that have led to the deindustrialization of the American economy and the decline of American cities include:

—the multinationalization of corporations, particularly American corporations;

—a new transnational economic order resulting from the postwar restoration of the European and Japanese economies and new industrial development of the Third World nations; and

—the economic shortfalls of the so-called postindustrial society.

Each of these changes is discussed here.

MULTINATIONALISM

One of the characteristics of the industrial economy is that it has always been global in the sense that the industrial establishments have crossed the boundaries of the nation-states reaching into far corners of the world in search of raw materials and markets for their manufactured goods. However, since World War II manufacturing firms, particularly American corporations, have begun to build factories in foreign countries to supply the local markets and to export products back into the U.S. market.

American corporations became multinational en masse in the 1960s and 1970s through establishing subsidiaries and developing joint ventures all around the world wherever they found hospitable political and social climates, abundant, stable, cheap, skilled labor, and opportunities for quick and large profits. This is the forceful attraction of the new transnational frontier.

The most important of the corporate policies of the American multinationals affecting the U.S. economy and its cities is the export of American capital and technology to foreign countries. This policy has directly affected investment policy for domestic operations; namely, negligence to renew old plants and to adopt advanced technology. It has inflicted double jeopardy on the American economy and American cities; first, it has helped expedite economic development in those countries receiving American capital and technology; and second, the neglect to invest in domestic plants and technological innovations has inflicted fatal wounds to the competitive capability of domestic operations in transnational markets.

Corporate leaders of American multinationals have pointed to other reasons for their industrial predicaments in the United States, such as: (1) excessive government regulations, (2) high labor costs, (3) the energy crisis, and (4) unfair competition from foreign producers.

What is apparent is that the economic plight of American cities and urban decline have been caused mainly by the deliberate policies of American corporations, policies of exporting capital and technology abroad, and the consequent disinvestment for domestic capital expansion and technological innovations. It is indeed frightening to realize that the failure of American corporations in domestic operations does not mean the failure of the firms as multinationals, but it certainly means a failure of the American economy and a decline of American cities.

TRANSNATIONAL COMPETITION

For two decades following the end of World War II, the United States enjoyed an economic supremacy challenged by no one in the world. Beginning in the early 1950s, however, the Western Europe countries and Japan recovered from war damages and rebuilt their industries. In the 1970s the Third World developing nations began to develop new industrial capability and enter into competition in the transnational markets. Intended or not, the corporate policies of the American multinationals should be given a lion's share of credit for their contributions to the process of rebuilding European and Japanese industries and stimulating the new industrial developments of the Third World nations through the transfer of American capital and technology. However, this emphasis on the role of American multinationals in the economic development of other countries in the world is not intended to underrate the decisive roles played by business, government, and labor of the countries affected. It may be fair to say that the contributions made by the American multinationals to the economic development throughout the world have been much greater than those provided through the economic and technical aid policies of the American government in the immediate postwar era.

THE SHORTFALLS OF THE POSTINDUSTRIAL SOCIETY

In the postwar era, as employment in manufacturing lost in relative weight in comparison with employment in service industries, some social scientists saw these changes as the sign of a social and economic transition into a postindustrial society. Such a society is characterized by a high-technology economy backed up by knowledge and information that is heavily automated and highly productive, and the economy grows exponentially. The postindustrial economy, being knowledge and information intensive, is heavily engaged in information generating, information processing, and information disseminating as an important part of economic activity.

What is important in the message heralding the arrival of that postindustrial society is that the high tech and knowledge-intensive economy is supposed to replace the traditional assembly-line dominant manufacturing economy (Bell, 1973; Naisbitt, 1980).

Indeed, high technology fields are the bastion of American excellence. The United States is unrivaled in such areas as computer science, communication technology (including satellites and telecommunication), microprocessing, selected weapons, medical science, office machines, advanced research, and advanced education. However, the transition from industrial to postindustrial society has faltered, creating a serious employment shortfall in that the high technology and knowledge-based service economy has not grown sufficiently so that new employment in this sector equals or exceeds employment declines in manufacturing.

More importantly, the question in order here is whether or not the traditional industrial economy is replaceable by a high technology and advanced service economy in the United States. Would it be possible for the traditional industrial labor force to adjust to the postindustrial economy—and how well?

Japanese experience so far seems to suggest that fully automated factories succeed in retaining all of the displaced assembly line workers. Japan has a cultural tradition of regarding labor as a permanent part of a company and companies follow a policy of life-time employment. In addition, the Japanese labor force is so well educated and so highly trained that assembly-line workers, when relieved, can be reassigned to other functions such as planning, designing or marketing, for example (Lohr, 1981).

The type of transition being experienced in Japan may not be applicable to the United States when high technology is applied to industries. The semiskilled assembly-line workers here are most likely to be displaced and dislocated, most of them permanently. Education and training as provided by the U.S. public education system and public programs, as well as company efforts, are clearly inadequate to prepare the industrial labor force for technical and knowledge-based jobs in the post-industrial economy. Keeping unnecessary labor on the payroll is also against the American creed of good management because it is costly. Undoubtedly, technology is advanced enough to support a postindustrial economy in America, but the America labor force has not yet become postindustrial and American culture has not undergone the transformation necessary to become suitable for a postindustrial economy.

America once underwent a massive economic transition from agriculture to manufacturing. That transition did not replace agriculture with manufacturing, but rather added manufacturing to a continuing and strong agriculture base. In the current economic transition, the development of a high technology service economy should have been pursued as a way of adding a new dimension to the overall economy, not as a replacement for manufacturing. Because of this problem in economic transition, a deindustrialized economy may result instead of a new postindustrial economy and a postindustrial society. Simply put, a Silicon Valley cannot make up for the loss of many Detroits.

PROSPECTS FOR THE AMERICAN CITY

What is in store for declining American cities in the future? Can they be revitalized? Will they continue to decline? Will they achieve a relative stability at a certain point? These are certainly complex and difficult questions, and whatever answers one may come up with are bound to be uncertain. But one thing is clear and that is the future of American cities will be determined principally by the condition of the American economy. Thus, American cities and the American economy will share a common destiny in the future as they have in the past.

REINDUSTRIALIZATION

Reindustrializing and thus rebuilding the economy and revitalizing cities are the most desired developments by all parties concerned: government, business, labor, and the public. But can the United States reindustrialize to the extent that productivity, product quality, and prices can be made sufficiently competitive in both American and transnational markets to permit the recall of the displaced labor force?

For decades, U.S. government policies have encouraged corporate investment in capital innovations by offering increasingly generous tax incentives. During the Carter Administration, tax incentives were further enhanced to encourage business investment in declining cities through the provision of differential investment tax credits. The Reagan Administration has given the largest tax reduction to businesses and individuals so far to encourage savings and investment for business expansion. Local governments have actively pursued economic development for years. Particularly during the past several years, the so-called public-private partnership has been used as the new vehicle for local economic development efforts. So far the effects of these policies are hardly evident.

The success of Reaganomics in restoring the vitality of the American economy is dependent upon the willingness (1) of corporations to invest income created by tax cuts in renovating their domestic capital facilities and updating industrial technology, and (2) of individual taxpayers to save their additional disposable income so that it can be channeled into industrial and business investment.

The question is, will corporations in fact invest their tax savings to expand and renovate their industrial capital facilities? If the examples of recent corporate investment are indicative of future behavior, then the realization of this exception is doubtful.

The oil companies, which reaped enormous windfall profits in the years following the 1973 oil crisis, have often invested in purchases unrelated to energy development, notably in real estate. United States Steel Corporation's purchase of the Marathon Oil Company is another case demonstrating the lack of intent on the part of corporations to invest in order to rebuild their industrial machines.

Another area of critical importance is labor-management relations. There is no doubt that American corporations need dedicated and productive labor that will produce more at a lower cost, but present corporate management policy does not inspire loyalty and dedication from labor. Shrinking employment and plant closings are used as a whip to threaten labor and force it to agree to unfavorable and humiliating contract terms, thus seeding more hostility and resentment. A case in point is the recent agreement between the Timken Company and the United Steel Workers Union as a condition to build a $500 million new plant in a suburb of Canton, Ohio. A key term in the new contract is a ban on strikes in the next eleven years. The steel workers in Canton apparently agreed to the contract, not necessarily because of their loyalty and dedication to the company, but because the company threatened to build the plant elsewhere if the union failed to agree to the contract terms proposed by the company.

Drucker pinpoints the philosophical dilemma of American corporations' labor policy when he notes that labor is regarded as "cost" rather than "resource" (Drucker, 1982). Drucker outlines three approaches to reindustrialization: (1) automation, (2) redesigning the production process, and (3) application of microcomputers to production machines so as to self-monitor product quality in the production process. These three approaches all require adoption of advanced technology and conversion of the American industrial labor force from semiskilled labor to semiknowing white-coat technicians (Drucker, 1982).

Clearly, this challenge must be met by the corporations through a systematic reorientation of their corporate policies for capital investment and manpower development. Government, labor, and the public must support this transition, surely in ways different from the past.

THE POST-AFFLUENT SOCIETY

The postindustrial society predicted for America has actually turned out to be a deindustrialized society. The deindustrialized society is a post-affluent society. The economy in a post-affluent society is either in a state of decline, stagnation, or slow growth and "affluence is behind us" (Gappert, 1979).

The American dream of universal affluence is a failed dream in the post-affluent society where society becomes increasingly polarized between the rich and the poor, while the traditional middle class becomes an endangered species facing possible extinction. Those who are absorbed into the high-tech and knowledge-based service industries will enjoy high income and high status, while those who are not will become an underclass, perhaps a permanent one, for they will be nonfunctioning members of the society.

What will cities be like in the post-affluent society? They are likely to become the homes of the underclass, the growing dependent population, isolated from the rest of society. The deepening social cleavages may cause violence and social unrest. The urban riots of the mid-1960s were riots against promises not fully delivered, but the possible social unrest in a post-affluent society may become revolts against despair and hopelessness. If indeed the post-affluent society is besieged with underclass violence, it is inevitable that cities will become the highly garrisoned portions of a police state.

For those who manage the high-tech industries, the choice of location for business and home will be highly flexible because contact and communication will be unaffected by distance or location because of advanced communication technology. High-tech companies located in exurbia today, surrounded by lawns, trees and lakes, seemingly isolated from everything unpleasant, are an example of new locational patterns in the post-affluent society. Those who are trapped in the deindustrialized cities and those who are not will constitute two societies, separate and more unequal than ever before.

President Reagan has recently reminded Americans of that frontier spirit of "voting with your feet," implying that those who have lost jobs in the northern cities should move to the South and West to find new opportunities. The precise predicament of those who have failed to follow the new job trails to Houston or Dallas is that these displaced manufacturing workers lack marketable skills in the high-tech and service industries.

CONSEQUENCES OF FRONTIERISM

Frontierism or the concept of enhanced opportunities in a new place is one of the driving forces in American life. Frontierism has provided American society with several powerful benefits, including a heightened sense of upward mobility (a perception that has helped to minimize social conflict), a perception that the nation's store of resources and frontier settings is limitless, and the perception that the future (frequently imagined in a new setting) will be better than our present or past. Frontierism, on the other hand, has created many problems. Among the more important of these is the sense of impermanence built into the physical setting of our communities. Housing, neighborhoods, and even entire communities are imagined as having a finite existence. They will be used up and then discarded or filtered down to poorer and less socially accepted groups.

The major public policy issues facing American communities today are derived in great part from worn out frontiers. On an interregional scale, issues such as metropolitan coordination, differing levels of fiscal capacity among political jurisdictions, social issues, including racial segregation, centralization of the poor, variations in the quality of the housing stock, all were at one time or another created by frontierism.

Transnational frontierism creates an entirely new and highly intractable set of issues. Transnational frontierism fosters obsolescence of section of the national market perspective. Transnational markets and an international scale of operation favor the multinational firm. As discussed earlier, multinational firms are not constrained by political boundaries and, thus, seek opportunities on a broader geographical scale.

CHANGES IN LEADERSHIP ORIENTATION

Pareto (1902) asserted that the composition of leadership groups is constantly changing (the rise and fall of elites). What does remain stable in most communities is the privatistic, paternalistic, or civic-minded orientation of the leadership groups. New leadership elements may replace older and declining ones, but they tend to assume the prevailing leadership ethos. A change in leadership orientation from privatistic or paternalistic to civic-minded or vice versa occurs as a result of service socioeconomic shock. Existing leadership is found to be unable to deal with stressful conditions. The classic historical response to this inability of leadership to deal effectively with its constituency is revolution. In American communities revolution or sudden leadership change does not occur frequently. More often leadership change occurs when existing leadership has decided to seek a new frontier of opportunity elsewhere. The problems besetting older industrial towns in the industrial Northeast and Midwest are derived in large part from an unwillingness of the existing leadership to invest in the region. In such a setting new leadership orientations can replace existing or collapsing leadership orientations. Civic-minded leadership, under these conditions, can replace older leadership forms.

Privatism and an individualistic political culture have energized the urbanization of the American society and industrialization of the American economy is also the result of privatistic and individualistic excess in corporate policies and union strategies as well as the inability or unwillingness of the government to intervene or to control them.

As the New Deal was born in the midst of the economic consequences of the Great Depression, the deindustrialized and post-affluent society may usher in a new era of more complete social and economic equity in the distribution of limited wealth and opportunities.

What would possibly cause these changes in values and policies? The shock of social and economic catastrophe has already been discussed. When the social and economic cleavages escalate conflict and violence to a catastrophic proportion, a drastic change from an exclusionary society to an equitably sharing society may become acceptable. Political and public leaders may become change agents to promote social and economic equities instead of serving as the custodians of the status quo of the divided society.

NOTE

1. Looking to the future in America has always implied progress. Banfield develops his class hierarchy around the degree to which groups of people plan for the future. Middle- and upper-income people defer current consumption for a better future, while lower-income and "culture of poverty" people consume in the present with little regard for the future. Most Americans, however, look to the future for improved social and economic conditions.

REFERENCES

BANFIELD, E. (1974) The Unheavenly City Revisited. Boston: Little, Brown.
BELL, D. (1973) The Coming of Post-Industrial Society: A Venture in Social Forecasting. New York: Basic Books.
DRUCKER, P. (1982) The Changing World of the Executive. New York: Truman Tulley.
ELAZAR, D. J. (1970) Cities of the Prairie. New York: Basic books.
GAPPERT, G. (1979) Post-Affluent American: The Social Economy of the Future. New York: New Viewpoints.
HANTEN, E. W., M. J. KASOFF and F. S. REDBURN [eds.] (1980) New Directions for the Mature Metropolis: Policies and Strategies for Change. Cambridge, MA: Schenkman.
LOHR, S. (1981) "New in Japan: the manless factory." New York Times, December 13.
NAISBITT, J. (1980) "High tech/high touch: the restructuring of america." Presented at the Public Management Research Conference, Brookings Institute, Washington, D.C., November 17-18.
PARETO, V. (1902) Les Systemes Socialietes, Paris.
SOVERN, M. L. (1982) "The case for keeping U.S. aid to colleges." New York Times Magazine, February 7.
TURNER, F. J. (1920) The Frontier in American History. New York: Hold.

16

The Future of Urban Danger

SALLY ENGLE MERRY

☐ AMERICAN CITIES OF THE 1980s seem dangerous places, riddled with violent, unpredictable crime and persisting racial tensions. Will this be true of cities in the twenty-first century? Will they regain the relative peace and calm that crime statistics suggest was characteristic of the 1930s and 1940s, or will the crime explosion and mounting fear of crime of the postwar period continue unabated? An answer requires an understanding of the social forces that generate a fear of crime. If these forces persist, as it appears they will, so will public concern about urban danger.

The dangerous city is neither a peculiarly American phenomenon nor a particularly recent one. Concern about danger and the dangerous classes appears in large, heterogeneous, industrial cities in Europe, Africa, Asia, and Latin America and has accompanied the industrial growth of American and European cities (Brace, 1872; Chevalier, 1973; Tobias, 1967). Although preindustrial cities were often plagued with crime, they were apparently not viewed as particularly dangerous places (see Sjoberg, 1960; Miner, 1953).

We commonly assume that cities appear dangerous simply because of their high crime rates, but the more basic causes are found in the social structure of the large, heterogeneous, industrial city itself. In

AUTHOR'S NOTE: *This research was supported by a predoctoral dissertation research grant from the Center for Studies of Metropolitan Problems of the National Institute of Mental Health 1 F31 MN05088-01. I am grateful for the perceptive comments of Joan Bamberger, Peter Hainer, Robert Hunt, David Jacobson, Jenny Phillips, and Susan Silbey on earlier drafts of this chapter. Lisanne Crowley contributed greatly to the research on cross-cultural and historical aspects of urban danger.*

order to understand why urbanites consider cities dangerous, and why some do so more than others, I spent a year and a half doing anthropological field research in a small housing project during 1975 and 1976 (see Merry, 1981). This community of 1200 people was afflicted with one of the highest crime rates in the city, a major eastern metropolis. In a victimization survey of two-thirds of the households, I found that almost half (48%) had experienced an assault, robbery, or burglary since moving into the project, while only one third (36%) had never experienced a crime of any kind. Although the residents did not know the actual crime rates, they were aware that they were living in a hazardous environment. I explored how they managed living with crime as well as their beliefs about what a crime is, what kinds of people are dangerous, how one identifies a dangerous person or place, and what cues signal a dangerous situation. Finally, I investigated how these cultural constructs were related to the social organization of the neighborhood.

A HIGH-CRIME NEIGHBORHOOD

Dover Square is the pseudonym for a neighborhood of 1200 individuals tucked between Chinatown and a decaying white, black, and Hispanic neighborhood undergoing spotty gentrification. Its three hundred families reflect the ethnic diversity of the surrounding area: 52% are Chinese, 27% are black, 12% white, and 6% Hispanic. The project was opened in 1966. Yet, after ten years of coresidence, these diverse ethnic groups have not blended into a single community. Instead, each maintains a separate social world, a network of friendship, kinship, and communication links that stretches across the project uniting those of the same ethnicity while separating neighbors of different ethnic groups. The groups were thoroughly mixed when the project was rented, and the ensuing years have not produced any ethnic clusters or enclaves in the development.

Each group is an immigrant population and brings with it distinctive cultural and social structural characteristics. The Chinese residents are generally China-born immigrants, two-thirds of whom came from rural peasant villages during the 1950s and 1960s and one third of whom fled crowded Hong Kong in the 1970s. Of the men, 86% work in Chinatown or the suburbs as cooks or waiters in Chinese restaurants or in Chinese shops and grocery stores. Of the 65% of the wives who work, fully 90% are employed as stitchers in garment factories downtown or in Chinatown. Their English-speaking children, however, pursue American educational ideals and many attend elite colleges and aspire to careers in engineering, science, and medicine. Those who do not succeed in school sometimes sink into a Chinatown underworld of petty crime and gang activities. Only 1% of the families are on welfare.

Most of the blacks migrated to the North during the 1950s and 1960s. They now constitute the occupational elite of the project. More work in lower-middle-class occupations than in any other group (about 12%), while another 26% are skilled manual workers; 25% are supported by welfare. About three-fourths of the black families have relatively stable jobs and are committed to values stressing work, church, and education for their children. The other fourth are families in which a bare subsistence is supplemented by a variety of illegal activities or in which family members have a past of petty crime.

The most conspicuous residents of the project are a group of about twenty young men and women who regularly visit and talk in the playground or laundromat in the center of the project. These youths are widely blamed for the crime in the project, and by their own admission, are responsible for at least some of it. Of the group, at least ten have been arrested and sentenced at some point, and another eleven have been involved in prostitution or pimping. Although the group is mostly black, it represents only a small fraction of the black population of the project. The fifteen active black members come from only eight families, about 10% of the black families in the project. Of those black families with children, only one-third have children who are even friendly with this group. Two white youths who are also core members come from two different families. There are a few Hispanics and Chinese who are more marginal participants.

The white population is about half Syrian-Lebanese in ethnicity, most of whom are related to residents of a larger Syrian-Lebanese community located in the adjacent neighborhood. The rest are a potpourri of ethnic groups, including Irish, Italians, Greeks, Jews, and New England Yankees. Many in this group come from rural or suburban areas. The number of whites has been dropping slowly but steadily during the last ten years. Those white families remaining feel more and more isolated from the surrounding community. Those with resources have already left; those remaining are the elderly without children, young mothers on welfare, and Syrian-Lebanese families who enjoy the proximity to their relatives and friends. Of these families 24% are on welfare, and 50% have unskilled manual jobs, primarily in food or custodial services. The Hispanic families are primarily recent arrivals from Puerto Rico, attracted by the growing Hispanic community a few blocks away. All have children and most live in two-parent families. Almost half (42%) are on welfare.

ATTITUDES TOWARD DANGER

Although the rates of victimization of the blacks and the Chinese are approximately equal, the Chinese are far more concerned about crime than blacks. My victimization survey showed that 46% of Chinese households had experienced a robbery, burglary, or assault, while 53% of the black households had had the same experience. In both groups 14% had experienced such a crime more than once. Yet, Chinese residents typically are much more aware of danger than blacks, discuss incidents of crime more frequently, and carry out more precautions against crime. I interviewed 101 individuals, representative of the project population by age, sex, ethnicity, and length of residence categories, about their attitudes toward danger. When asked if they considered Dover Square a dangerous place to live, only 18% of Chinese respondents said that the project was not at all dangerous, while 65% of blacks felt that way. On the other hand, 30% of Chinese respondents said the project was fairly or very dangerous, while only 13% of the blacks expressed this view. Whites have the highest rates of victimization of any ethnic group, yet show attitudes toward danger that are strangely divided. While living in the project 63% had experienced a robbery, burglary, or assault, and 30% had this experience more than once. When asked if the project was dangerous, 44% of the whites interviewed said that it was very or fairly dangerous, yet 33% said it was not at all dangerous. Hispanics have a low victimization rate of only 21%, and all of the small number interviewed said that the project was not at all dangerous.

THE MEANING OF DANGER

THE STRANGER

What accounts for this discrepancy between the experience of crime and the perception of danger? The answer is rooted in the social structure of the project, in the frequency with which members of each ethnic group encounter strangers whom they believe are predatory and vicious and beyond their control. People who know each other do attack and steal from each other, but these incidents are viewed not with fear, but with anger. It is strangers who are feared, not family and friends, even when they are violent and abusive. Those who know the identities and reputations of the youths who lounge in the center of the project describe their environment as far less dangerous than those to whom these youths are strangers. Given the social structure of the project, blacks are far more likely to know who these youths are than

Chinese, and whites somewhat more likely than Chinese. When individuals are pooled without regard to ethnicity, the questionnaires reveal a significant correlation between perception of danger and knowledge of the identities of these youths. This correlation holds regardless of victimization except for those who have been victims of crimes three or more times, who feel more fearful even when they know who the youths are. It is the residents to whom these youths are strangers who find the project more frightening. As a young black resident told a young Chinese resident she met through my research, "To you, all the blacks are dangerous because you can't tell them apart, but to me they are George, Johnny, and Jamesy, and I know who to look out for and who will not bother me." The black woman felt very safe in Dover Square, while the Chinese woman was constantly cautious and aware of danger.

Strangers appear dangerous because they are inevitably unpredictable and uncontrollable. A stranger is a person known only as a member of a social category, lumped into that category on the basis of his age, sex, dress, demeanor, ethnicity, location, and speech habits. These categories develop in settings where interaction is casual and unstructured and participants must develop ways of ordering those they encounter. Each category contains assumptions about the way such kinds of people behave, but categories may be more or less accurate depending on the experience of the observer and his familiarity with the kinds of person he is encountering. Categories are generalizations, clusters of diverse and distinct individuals. The observer inevitably ignores some important internal variations, and may even lump quite different groups together if he is ignorant about their differences. A more socially distant and unfamiliar stranger is lumped into a larger, less accurate category than one who comes from a more familiar social world.

In Dover Square, residents group strangers from familiar ethnic groups into relatively sophisticated categories, while those who belong to more socially distant and culturally alien groups are lumped into grosser and less accurate categories. Chinese residents, for example, know few if any black residents, and are unable to distinguish the few active in crime from the vast majority of black project residents who do not commit crimes. Because social networks are circumscribed by ethnic boundaries, they are excluded from the flow of information in the black social network about who is committing crimes, what this person's family situation is, and what he is likely to do in the future. Several Chinese residents commented helplessly, "To me, all blacks look alike." One Chinese woman who knows no English, for example, is constantly plagued with the fear that the group of black youths who regularly gather on the steps in front of her apartment to visit and smoke are plotting to break into her apartment. Since she cannot understand what they are saying, nor does she have any idea what their reputations are, she does not realize that they are not connected to the youth group that commits crimes and are simply using her stairs as a social center. Nor can her friends, who are all Chinese, alleviate her fears, since they are also not part of a gossip network that could provide this information.

Black residents, in contrast, know the names and families of the youths who commit crimes in the project and understand why they act this way, even if they do not approve. Most know which youths are likely to rob them and which will not. One black couple, for example, said that the project was not very dangerous because only three or four project youths are involved in crime, and they know who they are and where they live. When one of the more notorious burglars was released from prison, this information quickly circulated through the black social network, tipping off the blacks to the need to be more cautious. Black residents generally feel that it is only outsiders, people who are strangers to them, who are dangerous. Blacks are also more sophisticated at classifying black strangers they meet than are Chinese residents. The black residents feel safer, not because they are less victimized, but because they confront fewer dangerous strangers and are better able to single out who is likely to be hazardous and what he or she is likely to do.

The white residents who know some of the local youths personally and are familiar with the life of crime and hustling from their own childhood in the Syrian-Lebanese community nearby feel far less concerned about crime than whites who grew up in homogeneous ethnic neighborhoods or suburban areas and know neither the identities of the local youths nor how to distinguish threatening strangers from harmless ones. The latter group is also isolated from gossip networks that could provide information about the youths committing at least some of the crime. They accumulate guns and fortify their houses with locks and bars, but do not feel free from danger despite this defensive strategy.

Knowing the personal identities of hazardous individuals aids controlling their behavior since it makes possible the implementation of a variety of sanctions such as violent retaliation, gossip, and using the police and the courts (Merry, 1979). The stranger is immune to all of these pressures, once he escapes, since he cannot be located again. The Chinese families that express the least concern about danger are those that have teenage sons who understand street life and know the youths involved in local crime. When the grandmother of one of these youths was robbed, for example, he immediately went to the home of a leader of the black youth group and told him angrily that if anything happened to his grandmother again, he would hold the leader personally responsible. The grandmother was not molested again. This youth's mother says that she considers the project a fairly safe place to live. A series of purse snatches by a few blacks finally elicited a massive controntation between Chinese youths and black youths in which the black ringleaders were severly beaten. Thus, knowledge of personal identities is critical to imposing sanctions through violent self-help. Those who know who is attacking them can also call the police and report an individual by name rather than by a vague description, can take that person to court, or can retaliate in a range of violent ways. Even when these tactics do not deter crime, they create a sense of control. Psychological experiments suggest that is is helplessness in the face of unpleasant experiences that creates anxiety and fear, and that this fear can be relieved by a sense of control, even when the perception of control is not real (Seligman, 1975). Even when a victim does nothing, the knowledge that he could do something to the offender appears to reduce his fear.

Those white residents who know the local youths similarly do not feel that their environment is particularly dangerous. One white woman, for example, felt quite confident that she would not be molested because her son told a black youth in the purse-snatching ring that if anything happened to his mother, he would beat him up. Another white man had his tires slashed, but after finding out from neighborhood children who was responsible, took a knife to the alleged culprit's tires. This young man did not find the project dangerous. A Chinese neighbor who also suffered lacerated tires was unable to track down the culprit, however, and he did view Dover Square as a dangerous place. Even if a person does not know the names of the local youth group but has a friend who does, he feels safer. Both a black adult and a Hispanic adult know none of the black youths, but feel quite safe because their teenage children do and they could identify the youths to them. The few Chinese families whose black neighbors offered to look out for them felt much safer than those who lacked this support. One Chinese man knows only a single black youth, but he is a leader of the black youth group. The Chinese man calls him whenever he has trouble with any of the young blacks. Each of these people feels safe because he has personal links to the group of youths held responsible for local crime, and through these links believes he has some means of controlling them. It is the faceless stranger, known neither by name nor location, who is the most difficult person to predict or control and the most dangerous.

ENCOUNTERS WITH STRANGE CULTURES

Some people appear dangerous simply because they are culturally very different and appear to be moving into and taking over a neighborhood, school, or organization. Because they are culturally distinct, such people threaten to undermine the social system, to weaken an order and cohesion based on homogeneity and stability. When large numbers of people with unfamiliar lifestyles and apparently bizarre and irrational forms of behavior move into an individual's social space, they threaten his orderly and manageable social world. They may also attack the social status he feels he has acquired on the basis of the social composition of his neighborhood. These people are strangers too, but in a different sense from the faceless criminal: They are people who may be known as individuals, but who inevitably remain outside the observer's social system. They are similar to Simmel's description of the stranger as someone who is in a group but not of it, a potential wanderer who comes and stays yet never gives up the freedom of leaving (1950: 402-408). Such people retain their sense of separateness and their distinctive way of life. This notion of the stranger is different from both the newcomer and the marginal man, two other common definitions of the stranger in the sociology of the stranger (see Levine, 1979). When many persons form a local social group that asserts its distinctive cultural identity, they constitute a stranger community (Shack and Skinner, 1979).

In Dover Square, whites feel that the Chinese residents are dangerous in this sense. They fear a takeover of the project and and discuss the dangers of increasing numbers of Chinese families moving in, which they think is turning Dover Square into Chinatown. Their comments frequently focus on objectionable features of Chinese life styles, such as their strange foods, customs of hanging ducks and chickens from the ceiling, and propensity to live with many people squeezed into a single apartment on "wall-to-wall mattresses." They rarely recognize the difference between the recent Hong Kong immigrants—struggling to make enough to support themselves and their families, accustomed to very dense housing in Hong Kong, and more closely involved with Chinese cultural patterns—and the more established Chinese families who have spent a generation becoming adjusted to American ways and raised very Americanized children.

Such families are rarely desperately poor and to my knowledge never had people other than family members living in their homes.

Chinese residents, on the other hand, find whites and blacks dangerous because they appear loud, aggressive, sexually immodest, and frightening when they handle conflicts by shouting and violence. Families that do not have a coresident husband and wife seem puzzling and disturbing, yet this is a feature of many black and white families that have single parents and often other relatives or friends in the home such as an uncle, grandmother, or lover. In contrast, the black and white families who share the values of a stable family, hard work, and commitment to education and church do not view one another as dangerous, nor do the blacks and the whites familiar with the street life find one another dangerous (Merry, 1980).

The notion that culturally distinct, unassimilated stranger communities are dangerous is a common theme in American history. Each wave of immigrants was viewed as more or less dangerous until it abandoned its distinctive way of life, language, and cultural traditions (see Higham, 1955). The Germans in colonial America, the slaves in colonial New York, the Irish in nineteenth-century Chicago, the Italians in early twentieth-century Chicago, and the blacks in the early twentieth-century South were each perceived as dangerous by the societies that surrounded them. The Chinese, who retained their ethnic exclusiveness longer and more extensively than many other immigrants because of unique conditions of work and immigration, elicited particularly strong visions of danger centering on their way of life and on their failure to blend (Miller, 1969).

As each immigrant group arrived, it was accused of moral laxity and criminality and perceived as a culturally and racially distinct entity that was unwilling to blend. During the mid-nineteenth century flood of Irish immigration to Boston, for example, one mayor of the city noted the common fear that the Irish are "a race that will never be infused into our own but on the contrary will always remain distinct and hostile" (Handlin, 1941: 185).

Similar fears appear in British and French urban hisotry (Beattie, 1974; Chevalier, 1973). Even in the Elizabethan era, Londoners disliked, mocked, and attacked foreigners (Salgado, 1977). By the nineteenth century, the Irish immigrants in London were merged with the paupers as the dangerous classes and viewed as shiftless, savage, degraded, disorganized, and a bad influence on the English working man despite the fact that the English worker lived about the same style of life as the Irish poor (Engles, 1845; Lees, 1969: 361).

MORAL DEVIANCE

Danger in Dover Square is also a threat to morality. Alcoholics slouched on street corners, people who conduct sexual activities on the street in full view of children, prostitutes on the sidewalks, transvestites parading in the playground, and well-heeled youths displaying their wealth and leisure earned from illegal activities in the center of the project all threaten the moral order. The danger is particularly acute for black and white parents who are trying to steer their children away from the street life and keep them in school or at a job. The ringleaders of this group are described as dangerous by working-class parents, not because of personal threats but because they tempt their daughters into prostitution and their sons into street crime. They are less of a threat to Chinese parents since the youths are outside their social world. Chinatown gang life and martial arts clubs pose an analogous threat to Chinese parents, however, who often describe Chinatown as dangerous. Uncollected trash, streets that are not cleaned, broken fences, and a general state of disrepair are also perceived as a kind of danger since they suggest that the community is weak, that nobody cares about the neighborhood, and that they are easy prey. Even gossip, which attacks only an individual's moral reputation, is often cited as a major danger by those whose criminal activities or illicit love affairs form enticing subjects for conversation.

The city has long appeared as a place of vice, immorality, and corruption, and early discussion of the dangerous city focused on its threat to virtue. A nineteenth-century treatise on the dangerous classes of New York blends vice and crime, describing the group as "a great multitude of ignorant, untrained, passionate, irreligious boys and young men" (Brace, 1872: 28). These classes are dangerous because they commit crimes, threaten the moral order, and are potentially revolutionary (Brace, 1872). An early twentieth-century study of Boston describes the dangers of the lodging house system: Because there are no parlors, young people are forced to visit in their bedrooms, with clear hazards to chastity (Wolfe, 1913). The population of single women in colonial New York evoked fear and distrust since it was suspected that the women supported themselves illegally by engaging in prostitution, adultery, and otherwise disrupting family life. Much of the popular literature attacking the Chinese in the nineteenth century focused on their alleged sexual perversions, use of opium, cruelty, and immorality (Miller, 1969).

STATE COERCION AND COLLECTIVE VIOLENCE

A fourth and largely ignored meaning of danger is the threat of state control and collective violence generally exercised by dominant groups against subordinate groups. Many Dover Square residents see the police as abusive, arbitrary, unmanageable, and potentially violent, and encounters with police as humiliating and dangerous. The black youths claim that the police officers constantly stop them on foot or in cars asking for identification and explanations of their activities. The blacks believe that the police harass any local residents who are similar in race, sex, and place of residence to those who commit crimes. Residents believe that the largely white police force suspects all blacks and assumes that young blacks are all guilty of some criminal activity. One young man said that one of the most frightening moments of his life was being held at gun point in an isolated spot by a police officer who mistook him for a renowned criminal. A white woman claimed that the police are dangerous because they conduct high-speed chases through the courtyards and sidewalks of the project where children play and fire their guns in the dense residential area.

White policeman are often unfamiliar with the style and language of the local black teenagers and cope with confrontations with these youths with a blustery, aggressive, and demeaning assertion of their authority. I observed one white police officer attempting, at the request of the manager, to evict a local black youth from the project laundromat. The youth had used the laundromat as a social and recreational center for several years and felt that he belonged there. Unfamiliar with the area and the youth, the police officer burst in the door, demanded to know who was causing trouble, and belligerantly shouted at the youth that he had no business coming in and bothering people and must leave immediately. The youth, who had been quietly visiting with his friends, pretended to talk on the pay telephone and did not budge. As the weakness of the policeman's position as a lone, aging officer with neither a cruiser nor a partner became dramatically clear, he threatened to arrest the youth the next time he got into any trouble, issued a few insults, and left. Although the youth emerged victorious from this incident, such encounters, particularly in the presence of friends, are humilating and unpleasant. This incident reveals the extent of distrust and hostility between young blacks and white police officers, and contrasts dramatically with the ability of black security guards to achieve far more cooperation from the youths by adopting a more sympathetic and respectful style. Further, I heard accusations, which I was unable to verify, that officers unjustifiably beat young people during arrests, while held in custody, and while visiting friends held at the police station. Whether true or not, these accusations lend further credence to the belief that the police are violent, unpredictable, and dangerous.

Ironically, Chinese residents advocate more police as the solution to the crime problem, yet they are often unable to communicate with police officers. Police generally appear to be large, aggressive, loud, overbearing, and often insensitive to their problems. Many Chinese complain that when a police officer arrives after a burglary, he simply tells them that is is their fault "because you people keep too much money in your homes." The goods are rarely if ever returned or the culprit caught. Chinese families often feel that the insult and humiliation of contact with the police is not worth the meager chance that something will be returned. Furthermore, many Chinese residents fear to call the police because they think this will drag them into long court proceedings in which they will have to testify publicly against criminals and risk retribution as well as losing money from missed time at work.

The Chinese population in the United States has long regarded the courts as an unfamiliar and dangerous institution. During the era of restricted immigration (1882-1965), many immigrants came illegally and constantly feared detection and deportation. Even in the mid-1970s immigration officials appear in Dover Square tracking illegal immigrants. Chinatown leaders have long preferred to deal with internal problems of social control themselves instead of invoking American law enforcement. As a result, Chinese residents, particularly first-generation immigrants, shy away from American judicial institutions and regard them with some distrust and fear.

The sense of the danger of the state and dominant groups frequently appears among subordinate groups, particularly when the dominant groups regard them as dangerous and threatening. South African blacks and American blacks each perceived the law enforcement system of the state as dangerous. American history provides countless examples of both state control and collective violence (vigilantism, lynching, and massacres) against minority and stranger communities such as the Chinese, blacks, Catholic Irish, Italians, and numerous other groups. Intriguingly, Needham observes that the word for danger comes originally from the Latin *dominum*, which means the power of subjection the master has over his slave (1979: 44). Later, to be in someone's danger meant to be at his power or mercy (1979: 44). The dangerous classes were always the poor and the dominated (Brace, 1872; Monkkonen, 1975; Chevalier, 1973) who were themselves in danger from the dominant groups.

SUPERNATURAL DANGERS IN SMALL-SCALE SOCIETIES

This analysis of the meanings of danger bears intriguing similarities to notions of supernatural dangers in small-scale societies discussed by Douglas (1966) and Turner (1967). In her analysis of purity and pollution, Douglas argues that, in general, what is unclear and contradictory in society tends to be regarded as polluting, unclean, and possibly dangerous (1966). Persons who occupy transitional or ambiguous statuses, such as the unborn child or the boy undergoing a rite of passage, are thought to have the power to harm and to be particularly vulnerable to harm. Individuals who occupy ambiguous or weakly defined roles or operate in inarticulate parts of the social system are accused of possessing involuntary spiritual powers such as witchcraft. Those who occupy interstitial positions, operating in a social subsystem in which they are not full members but appear as intruders, are also thought to wield dangerous powers. Such people have a clearly defined status in one subsystem and an ambiguous one in another, as for example, the father in the Trobriand or Ashanti matrilineage or the mother's brother in the Tikopia or Tallensi patrilineage.

Douglas argues that outsiders appear dangerous in Western secular society as well as in simple, ritually ordered social systems. She cites the reluctance to accept back into society those who have been outside society in mental hospitals or prisons. They remain marginal and are considered unreliable and possessed of negative social attributes: The taint of being outsiders remains. The rest of society feels a need to protect itself from these dangerous individuals (1966: 117-118).

Turner develops this argument further in his discussion of the ritual state of liminality, the stage of transition between one state and another during which the individual does not occupy any status that is part of the hierarchical structure of society. Society is a structure of positions, while the period of liminality is an interstructural, or unstructured situation (1967: 93). This ritual state represents nonstructure, and consequently appears dangerous and anarchical to those concerned with maintaining the structure of society. Those in liminal states are often regarded as possessing dangerous, polluting, and inauspicious magico-religious powers. Turner distinguishes between two polluting, dangerous states: those which are caused by ambiguously or contradictorily defined statuses and those which derive from ritualized transitional states (1967: 97). The first are states defectively defined and ordered, the second are those which cannot be structured, such as liminal states.

Thus, according to Douglas and Turner, actors in small-scale societies seeking causes for their misfortunes often identify the locus of the threat as those who do not fit into the ordered hierarchy of society, those who are beyond its rules and sanctions and its power to order and control behavior. The residents of a high-crime neighborhood use the same principles to pinpoint the source of the hazards they face. They similarly perceive dangerous persons as those who are unfamiliar and unknown, those who are outside their predictable social world.

Clearly, there are vast differences between the organization of simple and complex societies and in the kinds of hazards their members encounter. The resident of a small-scale society is afraid of accidents, wasting diseases, political and economic defeat, and death, while the urbanite particularly fears violent attacks and neighborhood change. The causes of misfortune are also perceived differently. The member of a small-scale society believes that many of his misfortunes are caused by the action of supernatural forces propelled by another's will, while the urbanite sees his misfortunes as the undeserved consequences of random hostile acts committed by others. Yet, in both kinds of society actors suffer misfortunes but are uncertain where the final responsibility lies. They do not know exactly whom to blame, nor is there any obvious connection between perpetrator and misfortune. Consequently, they locate the source of dangers they face in persons who are not full members of their social worlds: individuals in liminal positions, those outside their social and cultural world, those who are anonymous. Such strangers always have the potential for appearing dangerous, and in a climate of concern about violent crime or supernatural misfortunes, it is these individuals who will be blamed.

THE FUTURE OF URBAN DANGER

Whether cities of the future will remain dangerous places depends not only on their crime rates, but also on their social structures: whether they will continue to be the refuge of new immigrant groups who bring with them different ways of life, whether this differentness will perpetuate separate social worlds and sharp social boundaries, and whether class stratification and cultural diversity will continue to foster moral diversity.

The shifting labor demands of industrial capitalism and growing world economic inequalities suggest that migration to the city will continue into the future, and with migration will come the creation of separate social worlds and the perpetuation of the social mosaic that is characteristic of contemporary cities. As long as sufficient crime occurs to generate anxiety about crime, such a social structure will guarantee that there are strangers to serve as the focus for this anxiety. Even moderate reductions in crime rates will not reduce urbanites' sense of danger any more than lowering the rate of disease will control fears of marginal persons in small-scale societies.

If future cities continue to be stratified by wealth and power and if the underclasses remain culturally and racially distinct groups, which appears likely, the forces that generate a public concern about danger and crime will persist. Crime will still be a useful rubric for describing the threat posed by a discontented and disenfranchised underclass, one that blames that group and at the same time justifies repressive measures directed against them. As this group expresses its anger in criminal acts as well as more direct political protests, the middle classes and elites will continue to label them as dangerous and criminal.

The fear of crime will remain a useful way of legitimating antagonisms based on social, cultural, and racial differentness. We will see periodic law and order campaigns in response to demands for social and racial justice. As the bottom segment of society becomes more affluent but still relatively poor and powerless, the recourse to crime as a mode of protest may increase, as has occurred in contemporary Sweden.

Since class and racial stratification seem highly likely to persist in American cities, so will the public rhetoric of urban danger and the threat of crime. Yet, the repressive measures this justifies will maintain the danger for the underclasses, who must continue to dread the control exercised by the state over their activities and the institutions through which this control is exercised, such as the police, the courts, and the prisons. Perhaps in smaller, more culturally and economically homogeneous cities there will be an escape from this scenario, but a sense of urban danger will undoubtedly haunt the large industrial cities of the twenty-first century as it has those of the twentieth century.

REFERENCES

BEATTIE, J. M. (1974) "The pattern of crime in England 1660-1800." Past and Present 62: 47-95

BRACE, C. L. (1872) The Dangerous Classes of New York and Twenty Years' Work Among Them. New York: Wynkoop and Hallenbeck.

BROWN, R. M. (1969) "The American vigilante tradition," in H. D. Graham and T. R. Gurr, (eds.) Violence in America: Historical and Comparative Perspectives II. Washington, DC: United States Government Printing Office.

CHEVALIER, L. (1973) The Laboring Classes and the Dangerous Classes in Paris During the First Half of the Nineteenth Century (F. Jellinek, trans.), New York: Howard Fertig.

DOUGLAS, M. (1966) Purity and Danger. Harmondsworth, England: Penguin.

ENGELS, F. (1845) The Condition of the Working Class in England (W. O. Henderson and W. H. Chaloner, trans. and ed., 1958) Palo Alto, CA: Stanford University Press.

HALLOWELL, A. I. (1941) "The social function of anxiety in a primitive society." American Sociological Review 7: 869-881.

LEES, L. H. (1969) "Patterns of lower-class life: Irish slum communities in nineteenth-century London," pp. 359-385 in S. Thernstrom and R. Sennet (eds.) Nineteenth-Century Cities. New Haven, CT: Yale University Press.

LEVINE, D. N. (1979) "Simmel at a distance: on the history and systematics of the sociology of the stranger," pp. 21-37 in W. A. Shack and E. P. Skinner (eds.) Strangers in African Societies. Berkeley: University of California Press.

MAYER, P. (1961) Townsmen and Tribesmen. London: Oxford University Press.

MERRY, S. E. (1981) Urban Danger: Life in a Neighborhood of Strangers. Philadelphia, PA: Temple University Press.

——— (1980) "Racial integration in an urban neighborhood. The social organization of strangers." Human Organization 39: 59-69

——— (1979) "Going to court: strategies of dispute management in an American urban neighborhood." Law and Society Review 13: 891-925.

MINER, H. (1953) The Primitive City of Timbuctoo. Princeton, NJ: Princeton University Press.

MONKKONEN, E. H. (1975) The Dangerous Classes: Crime and Poverty in Columbus, Ohio 1860-1885. Cambridge: Harvard University Press.

NEEDHAM, R. (1979) Symbolic Classification. Santa Monica, CA: Goodyear.

SALGADO, G. (1977) The Elizabethan Underworld. London: J. M. Dent.

SELIGMAN, M. E. P. (1975) Helplessness: On Depression, Development and Death. San Francisco: W. H. Freeman.

SHACK, W. A. and E. P. SKINNER [eds.] (1979) Strangers in African Societies. Berkeley: University of California Press.

SIMMEL, G. (1950) "The Stranger, "pp. 402-408 in K. H. Wolff (ed. and trans.) The Sociology of Georg Simmel. New York: Free Press.

SJOBERG, G. (1960) The Preindustrial City. New York: Free Press.

TURNER, V. W. (1967) The Forest of Symbols. Ithaca, NY: Cornell University Press.

WOLFE, A. B. (1913) The Lodging House Problem in Boston. Cambridge: Harvard University Press.

Part V

Policies of Transition:
Can Cities Manage Their Transformation?

☐ THE FEDERAL SYSTEM of intergovernmental relations has been in a state of flux since Richard Nixon borrowed the concept of revenue sharing from the Brookings Institute in the early 1970s. Fifteen years later the significant issue remains: Do the states, cities and localities have the political will and the organizational competence to manage the developmental and distributional functions initiated by the federal governments over forty years of New Deal-Great Society legislation?

In their contribution Alm and Burkhead analyze the newest federalism against a historical and theoretical background. They conclude that for most cities, a fiscal safety net is not yet in place and the outlook may be "most dismal." Bickford and Vehorn look at the ways by which cities are likely to respond to cutbacks and shortfalls. They identify four likely consequences of reduced public services: innovations, lower expectations, increased disparities, and new private-public relationships. In the last chapter an epilogue on urban management suggests the need for (1) a new philosophy of urban development, (2) new management skills, and (3) a complex urban futures management model.

Some cities at least will be able to achieve an intentional and not an accidental transformation as they enter the twenty-first century.

17

The Future of Fiscal Federalism

JAMES ALM
JESSE BURKHEAD

> This massive federal grantmaking system has distorted state and local
> decisions and usurped state and local functions. I propose that over the
> coming years we clean up this mess. I am proposing a major effort to
> restore American federalism [Budget of the U.S. Government, 1982].

☐ THIS IS AN AWKWARD TIME to speculate about the future of
fiscal federalism in the United States. If the Reagan Administration
has its way, the federal system may be on the verge of a major transformation. The administration has already made significant changes
in the amounts and the methods by which federal grants are
distributed among state and local governments. More importantly, the
administration has outlined a ten-year plan aimed at transferring a
larger share of the responsibilities for public services to the state and
local level. However, the details of this realignment are yet to be
revealed, and specific proposals will not reach the Congress for some
time. Moreover, congressional reaction, as well as that of governors
and mayors, is most uncertain. Major issues are not likely to be
resolved before the 1982 elections, if then.

Thus the best that can be done here is to examine the Reagan
proposals, to recount the recent history of fiscal federalism, to look at
the economics of federalism, and to speculate about the resulting
fiscal condition of the cities. The main themes running through all of
this are two. First, American federalism is in disarray. The almost
irresistible tendency to intergovernmentalize all public issues has
meant that the most national as well as the most local concerns are
fought out in the intergovernmental arena. The result, in the words of

the Advisory Commission on Intergovernmental Relations, is that "intergovernmental relations have become more pervasive, more intrusive, more unmanageable, more ineffective, more costly, and above all, more unaccountable" (1981). Reform is clearly necessary. Second, however, the Reagan initiatives, though laudable in their intentions, are unlikely to provide these reforms. Cutbacks already enacted in federal grants have imposed severe fiscal burdens on state and local governments, but the hoped-for reduction in federal strings has not yet occurred. Social welfare programs that were thought to be in the national interest are now to be transferred to the states and cities but without adequate resources to maintain their levels. In short, neither the proposed realignment of fiscal responsibilities nor the change in the grant system is likely to improve the performance of the federal system, and both are apt to impose severe burdens on the cities.

THE BACKDROP

The economic experience of the past decade, now extending into the 1980s, has not been pleasant. The United States, like nearly all advanced industrial countries, has experienced rapid inflation, rising unemployment, lowered productivity improvement, and slow growth. Governments at all levels have been in a fiscal straitjacket, and fiscal fragility in the public sector has been matched by fiscal fragility in the private sector. The halcyon days of steady growth in real income, matched by an increase in public sector activities, is clearly at an end. There is no light at the end of the tunnel; indeed, the length of the tunnel cannot be perceived.

The response of national governments to this sorry state of affairs has, of course, varied greatly. In the United States the Reagan Administration has embraced supply-side economics. On the tax side, tax reductions extending through 1985 are to encourage incentives to work, to save, and to invest; additional investment will increase productivity, restore growth rates, provide jobs, and lower inflation. Federal expenditure priorities will also be reordered: Major increases in defense (war) spending will permit the United States to keep up with or surpass the Russians; federal spending on unproductive social welfare programs will be reduced, with many programs returned to the states; and federal grants to state and local governments will be consolidated and reduced. As our colleague Roy Bahl has stated, these latter policies constitute the "Newest Federalism" (1982).

The main features of the Newest Federalism—some of which are in place, some of which are only proposed—can be described as follows:

Phase I. A major restructuring of federal-state-local fiscal relations occurred in August 1981 as Congress adopted, under heavy pressure from the Administration, the Omnibus Budget Reconciliation Act (P.L. 97-35).[1] The Administration had proposed to consolidate 116 categorical grants into 15 block grants under state supervision. The Congress accepted 57 of these proposals for consolidation into nine block grants.[2] The states were required to assume responsibility for two of these by October 1, 1981—low income housing energy assistance and community services, and social services, including Title XX. The remaining seven must be assumed by October 1, 1982 for such programs as preventive health, maternal and child health, alcohol, drug abuse and mental health, and education. Federal grantor agencies have not yet published proposed regulations for all of these. The states are working out administrative arrangements to handle the blocks; however, many states, such as New York and California, are moving very slowly to establish new administrative machinery in the midst of uncertainty about funding levels.

Phase IIA. In the 1982 State of the Union Message the President proposed the return of some forty additional grant programs to the states and "the Big Swap"—Medicaid to the federal government, food stamps and Aid to Families with Dependent Children (AFDC) to the States (New York *Times*, 1982d). Some further detail was explained in the budget (Budget Special Analysis H, Fiscal Year 1983, 1982) The turnback programs have an estimated cost in fiscal 1984 of $30.2 billion and include such grant programs as water and sewers, child nutrition, alcohol and drug abuse, legal services, airport construction, urban mass transit construction, vocational rehabilitation, and General Revenue Sharing. The swap involves a current exchange of $20 billion; with Medicaid costs expected to rise more rapidly than AFDC-food stamp costs, the Administration projects an annual net saving for the states reaching $8 billion in fiscal 1987. The turnback and the swap are proposed to begin in fiscal 1984, with a special transition fund available to the states in the amount of $28 billion annually. To this fund will be dedicated the oil windfall profits tax and some existing excise tax receipts from gasoline, tobacco, alcohol, and telephone services. The states will share in the fund in accordance with the transfers they now receive from the forty categoricals. Some programs will require pass-throughs to local governments. In addition, there would be seven new consolidated grants and additions to the existing block grants (Special Analysis H, 1982).

Phase IIB Beginning in fiscal 1988 the Administration proposes that the federal government phase down the transition fund at the rate of 25% a year—the excise taxes will be reduced at that rate through 1991. Subsequently, the states would have the option of reimposing the excise taxes. By fiscal 1992 the federal excise taxes will expire and the states will have the option of "continuing, changing, or terminating more than 40 turnback programs...or increasing their tax base by substituting local taxes for federal taxes that will be phased out" (Budget, Fiscal Year 1983, 1982). The Administration's proposals for turnbacks and block grants are accompanied by an on-going program both to reduce the number and scope of mandates that have traditionally accompanied the grants and to reexamine the cross-cutting requirements. These latter have required state and local grantees to conform with statutes that require provision for the handicapped, equal employment opportunity, affirmative action, Bacon-Davis, and many others. The total of the cross-cutting requirements is usually put at over fifty.

Phases IIA and IIB are, of course, much more uncertain than Phase I, which has been enacted if not yet administered. Not only are Phases IIA and IIB complicated by unknown Congressional reaction, their future is also clouded by the attitudes of the governors and mayors. The Administration would apparently like to separate the Newest Federalism from budget policy; the governors in particular would like to look at both together (New York *Times*, 1982d,e). Moreover, the traditional position of the National Governors Association supports a transfer of all major income maintenance programs to the federal government. The Administration apparently regards gubernatorial support as important; negotiations continue (New York *Times*, 1982f).

THE RECENT HISTORY OF FEDERAL ASSISTANCE

There are few who would defend the intergovernmental system that President Reagan inherited. Indeed, over a twenty-year period it had become very nearly a nonsystem, overloaded, unresponsive, and in need of decongestion. The major factor in this change was a massive expansion in the role of the federal government. In 1960, for example, there were approximately 130 federal grant-in-aids that distributed $7.2 billion in aid primarily to state governments. In contrast the Office of Management and Budget (OMB) counted in 1980 more than 500 grant-in-aids and another 600 federal assistance programs, the latter including loans, loan guarantees, subsidized loans, surplus property transfers, and the like. State and local governments are more than ever dependent on federal aid; in turn, the federal government, if not actually delivering local public services, exerts more influence here than ever before, through its disbursement of revenues and its controls on the use of these revenues.

The Great Society of Lyndon Johnson was the launching pad for this transformation. During LBJ's administration the national interest was redefined and expanded to include major programs in such areas as health care, transportation, water and sewerage, urban redevelopment, crime control, and open space preservation. All of the Great Society programs, with the prominent exception of Medicare, were administered by means of grants-in-aid. The number of grants increased from fiscal 1963 through fiscal 1969 from 160 to 581 (Advisory Commission on Intergovernmental Relations [ACIR], 1977a).

The Johnson Administration did initiate a number of reforms to simplify the categorical grant system that it had put in place. In 1966, at the request of the president, Congress enacted the Comprehensive Health Planning and Public Health Services Amendments (P.L. 88-749), the first block grant, which came to be known as Partnership for Health (ACIR, 1977b). In 1968 a new program was enacted as a block grant—The Omnibus Crime Control and Safe Streets Act (P.L. 90-351) (ACIR, 1970, 1977c). In addition, the Johnson Administration supported the Intergovernmental Cooperation Act of 1968 (P.L. 90-577), with the objective of rationalizing local area review of grant applications. (This was later implemented by OMB Circular A-95.) Yet despite these reforms the Johnson presidency had no consistent philosophy about fiscal federalism. Great Society programs marched under the banner of "creative federalism"; there was to be a synergistic alliance of the federal, state, and local sectors with participation by not-for-profit organizations at the local level and occasional participation by private firms.[3]

The Nixon Administration entered office with the goals of simplifying the intergovernmental grant system and returning powers to subnational governments. Calling for a "New Federalism," the administration proposed the adoption of General Revenue Sharing and the consolidation of 129 categoricals into six block grants—education, law enforcement, rural community development, urban development, manpower, and transportation. These proposals met with only partial success. General Revenue Sharing was adopted in 1972 (P.L. 95-512). Manpower training became Comprehensive Education and Training Act (CETA) in 1973 (P.L. 93-203), one of the most complicated grants-in-aid in our history. Urban development became Community Development Block Grant in 1974 (P.L. 93-383). Rural development was put into a block by executive order. However, the education and transportation block grant proposals were defeated. In addition, Social Services (Title XX) (P.L. 92-603) for the indigent elderly, blind, and disabled was federalized in 1972. Additional categorical grants were added during the Nixon Administration.

President Ford continued in the spirit of New Federalism by proposing three block grants—in health, education, and child nutrition. The proposals were defeated. Some 30 new categoricals were legislated in this administration. During his tenure President Carter proposed additional grant consolidations in seven areas. Only one of these was adopted, and that in part: 44 categoricals in transportation were cut to 9 (P.L. 95-599).

The transformation in intergovernmental fiscal relations between FY 1960 and FY 1982 is indeed startling, as Table 17.1 exhibits. The increase in total federal grants-in-aid in these years was twelvefold in nominal dollars. Transfer payments intended for individuals accounted for a slightly larger portion of this increase, and transfers to the state-local sector a slightly smaller portion. Federal grants in relation to budget outlays were 7.6% in FY 1960 and peaked at 17.4% in FY 1978. State and local governments have become increasingly dependent on federal grants that have amounted to one quarter of their outlays in recent years, with large city dependency often exceeding 50%. And these figures mask the depth of federal control. Additional federal regulations, in such areas as the environment, health and safety, consumer protection, discrimination, and social equality, limit even more the discretion of state and local governments.

If the projections hold, the Newest Federalism will reduce federal grants as a proportion of budget outlays from the high point of 17.4% in FY 1978 to 9.6% in FY 1985. All of this reduction will be at the expense of the state-local sector; transfers to individuals are projected to increase slightly. However, the triad of grant types that has been in place since 1972 will surely continue, although with reduced funding: General Revenue Sharing, block grants, and categoricals. The number of categoricals will be reduced but many will remain.

THE ECONOMICS OF FISCAL FEDERALISM

The Reagan Administration's programs, both those already enacted and those only proposed, represent a dramatic change—in many respects, a reversal—in the direction of the American federal system. But is this turnaround desirable? It was suggested above that it is not.

Table 17.1
HISTORICAL TREND OF FEDERAL GRANT-IN-AID OUTLAYS
(Fiscal years, dollar amounts in millions)

| | Total Grants-in-aid | Composition of grants-in-aid | | Federal grants as a percentage of | | |
| | | Grants for payments to individuals[1] | Other | Budget outlays | | State and local expenditures[3] |
				Total	Domestic[2]	
Five-Year intervals:						
1960	7,020	2,479	4,541	7.6%	15.9%	14.7%
1965	10,904	3,931	6,972	9.2	16.5	15.3
1970	24,014	9,023	14,991	12.3	21.3	19.2
1975	49,834	17,441	32,392	15.4	21.5	23.0
Annually:						
1976	59,093	21,023	38,070	16.2	21.9	24.2
1977	68,414	23,860	44,555	17.1	22.9	25.9
1978	77,889	25,981	51,908	17.4	23.1	26.8
1979	82,858	28,765	54,093	16.9	22.5	26.1
1980	91,472	34,174	57,298	15.9	21.2	26.3
1981	94,762	39,934	54,828	14.4	19.5	25.3
1982 estimate	91,220	41,500	49,720	12.6	17.3	NA
1983 estimate	81,418	37,563	43,855	10.7	15.5	NA
1984 estimate[4]	81,853	39,951	41,902	10.2	15.1	NA
1985 estimate[4]	83,517	42,310	40,207	9.6	14.8	NA

[1] Income security and Veterans Administration medical care.

[2] Excludes outlays for the national defense and international affairs functions.

[3] As defined in the national income and product accounts.

[4] Data for 1984 and 1985 are included to be consistent with the multiyear planning system. They have not received as much review as the estimates for 1982 and 1983.

NA = Not available.

SOURCE: Special Analysis H, Fiscal Year 1983, Table H-6, p. 17.

However, this judgement requires some justification, and it is in the economic theories of fiscal federalism that the bases for appraisal have to be found.

There are two main issues in the fiscal federalism literature. The first concerns the appropriate level of government to which a specific governmental function should be assigned. The second focuses on those intergovernmental mechanisms that can be used to better perform these functions. Since the Newest Federalism has implications for both issues, each is considered in turn.

As suggested by Musgrave, there are three major functions that must be performed by the public sector: the stabilization, the distribution, and the allocation functions (1959). The issue that is of interest here is the appropriate assignment of responsibilities for these functions among the various levels of government. There is widespread agreement that the stabilization function should be performed primarily by the federal government (Engerman, 1965; Oates, 1968; Stern, 1973). The potential of leakages from state and local government spending, the burden created by government debt issued to citizens of other jurisdictions, the difficulty of independent monetary policies by subnational governments, the problems of coordinating many independent monetary and fiscal policies—all of these factors make it unlikely that an effective stabilization policy can be pursued at the state-local levels.

The attempt by a state or local government to redistribute income among its citizens is also unlikely to meet with much success. As shown by the exodus of higher-income individuals from the central cities to the suburbs and by the movement of businesses from high-tax to low-tax jurisdictions, there is substantial mobility of tax bases in the face of tax differentials. In sum, it is widely concluded that income redistribution must also be performed by the federal government (Stigler, 1957; Breton and Scott, 1978; Pauly, 1973).

The assignment of the allocation function is much more unsettled, and there are in general no clear advantages to either centralization or decentralization of this activity. The case for federal (or state) provision is strongest when there are public goods whose benefits are national (or regional) in scope, and when there are significant economies of scale in the production of the good. Examples of such goods at the national level are defense and education; at the state level, pollution control, waste disposal, and mass transit might be included.

However, it is unlikely that complete federal provision of public goods is always efficient. Centralized provision is apt to lead to uniformity across all communities, and, where the benefits of a particular public good are spatially limited to a specific locality, local government provision at levels different from neighboring jurisdictions allows closer adaptation of production to the demands of citizens. As formalized by Oates in his "decentralization theorem," a public good is most efficiently provided by that level of government whose boundaries coincide with the spatial distribution of benefits of that good (Oates, 1972; Musgrave, 1959; Breton, 1965; Netzer, 1968; Olson, 1969). The federal system, therefore, emerges here as the optimal form of government: "We can envision a system consisting of a multiplicity of government levels, where each government is responsible for providing the efficient level of output of the public good consumed collectively by the individuals in its jurisdiction... and this would lead to a welfare optimum (Oates, 1972).

Decentralized provision of local public goods may be beneficial for other reasons as well. If local governments provide different fiscal packages, a citizen may choose that bundle that best satisfies his tastes by "voting with his feet," although this option is not viable for all income classes (Tiebout, 1956). Moreover, decentralization may lead to greater accountability for state and local government decision-making and greater efficiency in government choices, since citizens may have greater access to, control of, and involvement with governments that are closer to them. Finally, in a system with many diverse state and local governments, competition among these governments may lead to greater experimentation and innovation in the provision of government services.

Economic criteria suggest, therefore, that there are clear gains to be had from a federal system of governments in which functional responsibilities are appropriately assigned: The central government should provide for stabilization and distribution, as well as for those public goods whose benefits are national in scope; state and local governments should supply those public goods that benefit only their citizens.

Equally clear, however, are the problems that may result when the full assignment of duties is made in this way. For one thing, government actions in one jurisdiction may generate benefits or costs that spill over into other jurisdictions. Since the government ignores these externalities, it will not choose an efficient level of production. Of perhaps greater significance is the lack of correspondence between the ability of a State or local government to handle a set of functions and its ability to finance them. A government is often unable to make use of tax bases that are seemingly available to it, due to legal limitations on its taxes or to fears of losing citizens and firms. In addition, there is great variation in the fiscal capacity of States, and even greater variation in that of cities. A poor jurisdiction may therefore be unable to provide even minimum levels of essential goods and services, except by imposing taxes at very high levels.

Happily, the existence of some forms of intergovernmental fiscal transactions makes the full assignment of fiscal duties to a single level of government unnecessary. By splitting up the responsibilities for the design, production, and finance of a public good among the governments, these mechanisms make it possible to exploit the advantages of the governments in performing specific functions and so to retain the benefits of a federal system. It is the search for the optimal form of these mechanisms that constitutes the second main strand of the fiscal federalism literature, and the instrument that has received the most attention here is the intergovernmental grant.

What types of grants are typically suggested (Breton, 1965; Oates, 1968, 1972; Schultz, 1974; Break, 1980)? Since the necessity for grants stems from inequities and inefficiencies in state and local provision of public services, grants should be used to address these problems. When interjurisdictional externalities are present, a conditional open-ended matching grant should be provided, with the matching subsidy equal to the value at the margin of the externality. When there are differences among jurisdictions in income and wealth, an unconditional lump-sum grant should be extended to those jurisdictions with significant fiscal deficiencies.

This is what the economics of fiscal federalism says about the structure of an efficient and equitable federal system. How does the Newest Federalism relate to these criteria?

JUDGING THE POLICIES OF THE NEWEST FEDERALISM

The Newest Federalism of Ronald Reagan seeks two broad goals: a realignment of fiscal responsibilities among the various levels of government and a simplification of the intergovernmental grant system. On the basis of the criteria outlined above, will these changes make American federalism more workable?

Consider first the realignment. There can be little doubt that some turnback of responsibilities to state and local governments is long overdue. There is now a maze of federal programs that serve no obvious national purpose but that simply reflect the substitution of congressional judgement for that of state and local officials. Federal programs for rat control, jellyfish control, snow removal, pothole repair, home insulation, bikeways, urban gardening, noise control—none of these meet the basic test suggested earlier as activities that generate substantial spillovers.

Such is not the case for "the Big Swap." Redistribution must be a central government responsibility. Unequal fiscal capacities among the states, combined with the movement of individuals and firms in response to tax differences, make state and local attempts to redistribute income both inefficient and inequitable. State provision of income maintenance programs would almost certainly create larger differences in support levels across states. And even if the states do maintain current benefit levels, these programs would be financed by taxes that are on balance more regressive than the federal taxes they replace. Poverty is a national concern, created largely by national economic policies. Only national programs can address its solution, and such programs should not single out as deserving only those poor who are old or sick or disabled.

In the area of grant reform, the basic Reagan strategy is aimed at the consolidation of the thicket of categorical grants into federal-state block grants. The intent here is fully consistent with the criteria for grant design suggested earlier. But the likely outcome is murky.

As noted above, the two block grants of the 1960s—Partnership for Health and Safe Streets—have been examined carefully by ACIR. The Community Development Block Grants program (CDBG) has also been well-researched, not only by ACIR but by others (ACIR, 1977d; Dommel et al., 1980; Dommel et al., 1978). The main lesson from these experiences is pessimistic: That which appears on the surface as a simple administrative choice between types of grants in fact involves a host of economic and partisan-political considerations, all intermeshed with the traditional conflicts that are inherent in multilevel government. One set of complexities is substituted for another (Burkhead and Birkhead, 1981).

The first consequence of a federal-state block grant is to transfer political and economic power from Washington to state capitals. As has often been noted by political scientists, the categorical grant is favored, almost as a pork barrel device, by Congressional authorization and appropriations subcommittees (see Beer, 1978). A categorical is supported by the "iron-triangle"—the coalition of interest group supporters, subcommittee members, and the grantor agency. Other pejoratives have been used to describe this triad of interests—"professional guilds" or "professional fiefdoms" (Seidman, 1980).

A "no-strings" transfer to the states may not survive the interests of the iron triangle over a period of time. The Partnership for Health experience is illustrative. Congress recategorized part of the block in 1970 and again in 1975. Safe Streets was subjected to the same treatment, earmarking within the block grant in 1971 and again in 1974. This kind of congressional behavior is hardly surprising in a society characterized by a host of interest groups, each supported by disparate funding sources. (The semantics are interesting. If one approves of this type of political pluralism, the lobbying activities are labeled "legitimate interest group influence." If one disapproves, the label is "special interests.") However, this behavior, which makes life on the banks of the Potomac so interesting, is serious, since some interest groups have much more political clout in Washington than in state capitals. The poor, the elderly, the Blacks may be particularly disadvantaged by the transfer of social programs from Washington to the states (Jacob, 1982).

Federal-state block grants also give rise to a set of political and administrative complexities.[4] In any one state the issue involves who shall make the resource allocation decisions within the block grant. Is it to be the governor and his budget office or a state department that has political and administrative experience—health, education, mental hygiene? What is the role of the legislature—should they or will they review the allocations within the block, or leave this to administrative discretion? Who shall determine how much of the block shall remain under state control and how much will be passed through to local jurisdictions or not-for-profit agencies? Will the iron triangle at the federal level be replaced by a series of little iron triangles in the fifty states?

Obviously all of this will differ from state to state. But there have been two developments over the past decade suggesting that legislatures will have more control over resource allocation than would have been the case fifteen years ago. Some state legislatures—Pennsylvania, Florida, Connecticut, and New York, for example—have asserted the authority to appropriate federal funds (New York *Times*, 1982f). And in almost all states legislatures are far better staffed than even a decade ago. These developments suggest that strong governors, accustomed to an acquiescent legislature in budget policy, may be in for some shock as legislatures insist on a major voice in the allocation of block grant funds[5].

There are other state-level difficulties with the blocks. In the last fifteen years many governments have greatly strengthened their administrative capacities. Professionalism has increased; central budget and planning agencies have been strengthened. Almost all states have become more involved and supportive of their local governments. To install state-level patterns to administer the block grants takes time, but a great many states are certainly possessed of the necessary ability.

Administrative competence is, unfortunately, not congruent with state fiscal strength. Although many states have strengthened their revenue systems, there is great variation in state fiscal capacities. The energy-rich states are in a particularly favored position at the moment. The block grants will provide these favored states with an opportunity to play catch up in their administrative procedures. More importantly, state administrative competence does not translate directly into sympathy for the fiscal plight of cities. It is not known how much of the federal-state block grants will be passed through to the cities. And even if the blocks are largely passed on, added flexibility at the state level may not mean greater discretion at the local level. Instead, state strings may simply replace federal strings.

Finally, planning problems are particularly acute when the federal government requires a planned use statement from the grantee, and even more acute when the planned use at the state level must reflect the intended use of resources at the local level—that is, when there must be an aggregate state plan that also reflects the intended use of pass-throughs. Partnership for Health and Safe Streets contained such state-local planning requirements. At present it is not at all clear whether the 1981 block grants or the proposed 1982 blocks will incorporate such provisions. Congress may not, constitutionally, appropriate funds without ascertaining that they are expended for the purposes intended (Article I, Section 9(7)). But how can block grants be financially audited when they are very largely fungible with state or local revenues? And how can performance auditing be accomplished without federal oversight, which is to be diminished? Similar problems have been encountered with the federal-local blocks—CDBG and CETA—to say nothing of the experience of the polar case of block grants—General Revenue Sharing (GRS).

CONCLUSIONS: THE URBAN OUTLOOK

The American federal system is most complex. The host of governments at all levels—nearly 80,000 at last count—makes it nearly impossible to delineate clear functional responsibilities and to implement a simple workable system of grants-in-aid. The Reagan Administration has attempted precisely these tasks. However, the Newest Federalism is unlikely to lead to a more efficient and equitable federal system.

As a first step in sorting out responsibilities, federal assumption of Medicaid and turnbacks of current federal programs with an obvious state-local focus are desirable. However, state takeover of AFDC and food stamps cannot but worsen the distribution of income. Similarly, simplification of the grant system should be pursued. However, block grants have so far failed to achieve their promise.

The impact of this transformation on the cities is particularly troublesome. In February 1982 the ACIR published a Staff Working Paper that attempted to analyze the turnback outcomes for ten or more packages of federal grant relinquishment and resource relief of the type proposed by the transition trust fund. All of the packages examined produced, of course, both winners and losers among states, with immense variations depending on the composition of the package. ACIR also analyzed a turnback package for twenty cities that would eliminate federal-local grants for CDBG, Urban Development Action Grants (UDAG), and Mass Transit, and substitute an equivalent amount of revenue sharing funds (ACIR, 1982). With one exception—Charleston, South Carolina—the cities would lose on the turnback, with Minneapolis the heaviest loser in the sample at 29% of own-source revenue and St. Louis not far behind at 21%. (Obviously counties and smaller communities would gain if replacement revenues are to be returned on the GRS formula.) If the states were required to replace revenue losses, the cities would suffer only until such provisions expired. However, the Newest Federalism does not yet contain such provisions. And if major cities are required to reduce service levels, are unable to replace aging infrastructures, or must search for a package of fees, service charges, and additions to the local property tax, the outlook is most dismal (Joint Economic Committee, 1981).

Some states will undoubtedly raise taxes, and the revenue may be made available to hard-pressed local jurisdictions. In 1981, interestingly enough, there were signs that the taxpayer revolt had abated a bit. More states raised taxes in that year than in any year since 1971—usually sales tax rates and liquor and tobacco excises (Public Administration Times, 1982). But even if this experience is replicated in 1982, there is no assurance that state-local revenue sharing will increase. In fact, there is some evidence that even the energy-rich states are not likely to respond to local jurisdictional needs (New York *Times*, 1982e). And many states have neither the political will nor the economic capacity to enact the major tax increases that are necessary.

To return to the starting point of this chapter, outcomes are most uncertain. What does appear likely is that the federal role will diminish in the years ahead, while the states are moved to center ring. Unfortunately, the cities are the acrobats in this circus; the fiscal safety net is not yet in place.

NOTES

1. For a discussion of the procedures and major features of the act see Keith (1981).

2. For an excellent brief summary of the grant consolidations and budget policy see *Survey of Current Business* (1981).

3. The ad hoc administrative policy of this period is set forth explicitly by Charles Schultze in testimony before the Ribicoff Subcommittee (1967).

4. See comments by Arthur J. Kremer, Chairman, New York State Assembly Ways and Means Committee in Burkhead and Birkhead (1981).

5. The support of the National Council of State Legislatures for the Reagan proposals is indicative.

REFERENCES

Advisory Commission on Intergovernmental Relations (1982) Changing the Federal Aid System: An Analysis of Alternative Resources/Responsibility Turnbacks and Program Trade-offs. Washington, DC: Government Printing Office.

——— (1981) An Agenda for American Federalism: Restoring Confidence and Competence. Washington, DC: Government Printing Office.

——— (1977a) Categorical Grants: Their Role and Design. Washington, DC: Government Printing Office.

——— (1977b) The Partnership for Health Act: Lessons from a Pioneering Blcok Grant. Washington, DC: Government Printing Office.

——— (1977c) Safe Streets Reconsidered: The Block Grant Experience 1968-1975, Parts A and B. Washington, DC: Government Printing Office.

——— (1977d) Community Development: The Workings of a Federal-Local Block Grant. Washington, DC: Government Printing Office.

——— (1970) Making the Safe Streets Act Work. Washington, DC: Government Printing Office.

BAHL, R. (1982) Testimony before the Joint Economic Committee, Washington, DC: February.

BEER, S. H. (1978) "Federalism, nationalism, and democracy in America." American Political Science Review (March): 9-21

BREAK, G. F. (1980) Financing Government in a Federal System. Washington, DC: Brookings Institution.

BRETON, A. (1965) "A Theory of government grants." Canadian Journal of Economic and Political Science (May): 175-187.

———— and A. SCOTT (1978) The Economic Constitution of Federal States. Toronto: University of Toronto Press.

Budget of the U. S. Government (1982) "The budget message of the president." Budget of the U.S. Government, Fiscal Year 1983. Washington, DC: Government Printing Office.

Budget Special Analysis H (1982) Fiscal Year 1983. Washington, DC: Government Printing Office.

BURKHEAD, J. and G. S. BIRKHEAD (1981) "Another crisis in fiscal federalism?" Ways and Means Report. Albany: New York State Assembly Ways and Means Committee.

DOMMEL, P. R., V. E. BACK, S. F. LIEBSHUTZ, L. S. RABINOWITZ, and Associates (1980) Targeting Community Development. Washington, DC: Government Printing Office.

DOMMEL, P. R., R. P. NATHAN, S. F. LIEBSHUTZ, M. T. WRIGHTSON, and Associates (1978) Community DEvelopment. U.S. Department of Housing and Urban Development. Washington, DC: Government Printing Office.

ENGERMAN, S. (1965) "Regional aspects of stabilization policy," in R. A. Musgrave (ed.) Essays in Fiscal Federalism. Washington, DC: Brookings Institution.

JACOB, J. F. (1982) Speech. Public Administration Times, February 1: 1.

Joint Economic Committee (1982) Emergency Interim Survey: Fiscal Condition of 48 Large Cities, 97th Congress, 1st session. Washington DC: Government Printing Office.

———— (1981) The Regional and Urban Impacts of the Administration's Budget and Tax Proposals, 97th Congress, 1st session. Washington, DC: Government Printing Office.

KEITH, R. A. (1981) "Budget reconciliation in 1981." Public Budgeting and Finance (Winter): 37-47.

MUSGRAVE, R. A. (1959) The Theory of Public Finance: A Study in Public Economy. New York: McGraw-Hill.

NETZER, D. (1968) "Federal, state, and local finance in a metropolitan context," in H. Perloff and L. Wingo (eds.) Issues in Urban Economics. Baltimore: Johns Hopkins Univ. Press.

New York Times (1982a) March 13: 1.

———— (1982b) March 7: 28.

———— (1982c) February 21: E4.

———— (1982d) January 27: A16.

———— (1982e) January 12: 1.

———— (1982f) January 7: 1.

OATES, W. E. (1972) Fiscal Federalism. New York: Harcourt, Brace, Jovanovich.

———— (1968) "The theory of public finance in a federal system." Canadian Journal of Economics (February): 37-54.

OLSON, M. (1969) "The principle of fiscal equivalence: the division of responsibilities among different levels of government." American Economic Review (May): 479-487.

PAULY, M. V. (1973) "Income redistribution as a local public good." Journal of Public Economics (February): 35-58.

Public Administration Times (1982) January 15: 2.

SCHULTZE, C. (1974) "Sorting out the social grant programs: an economist's criteria." American Economic Review (May): 181-189.

——— (1967) Testimony before the Ribicoff subcommittee on Executive Reorganization, 90th Congress, 1st session: 4300-4306.

SEIDMAN, H. (1980) Politics, Position, and Power. New York: Oxford University Press.

STERN, R. M. (1973) The Balance of Payments. Chicago: Aldine.

STIGLER, G. (1957) "Tenable range of functions of local government." Joint Economic Committee, Subcommittee on Fiscal Policy. Federal Expenditure Policy for Economic Growth and Stability. Washington, DC: Government Printing Office.

Survey of Current Business (1981) December: 21-24.

TIEBOUT, C. M. (1956) "A pure theory of local government expenditures." Journal of Political Economy (October): 416-424.

18

Changes in
the Provision of
Public Services

DEBORAH BICKFORD
CHARLES VEHORN

☐ RESOURCE SCARCITY IN AN ERA OF RETRENCHMENT
has created uncertainty over whether cities will be able to provide a
high quality of public services in the future (Blair and Nachmias,
1979). Coupled with this situation is the federal government's reduced
involvement in aiding local governments. Before this reduction, the
federal role was not clearly defined. It consisted of fragmented and
uncoordinated efforts at the programmatic level that led some to
conclude that the intergovernmental system is a directionless leviathan
(Brennan and Buchanan, 1979). Now there is a clear thrust to shift
responsibility back to the local level.

The primary purpose of this chapter is to examine factors that will
affect how adequately cities deliver public services in the future. A
special emphasis is given to the federal government's catalytic role in
changing the pool of resources available to cities. This past decade
marks a rapid change in local public sector activity. Many of the
forces affecting the performance of city governhents have been
outside the control of city officials, and have reduced their flexibility.
Thus we begin by reviewing major exogenous changes that constrain
city officials.

AUTHOR'S NOTE: *The opinions expressed by the authors in this chapter do not
reflect the opinions of the U.S. General Accounting Office.*

Next we address the federal role in city government, showing how this role evolved from a distant observer to an active, perhaps too active, helper. We follow this with a discussion of current initiatives at the federal level, such as the movement toward block grant consolidation, that will have an effect on local public service provision. The prospects for achieving a new balance in federalism depend in part on the success or failure of these political initiatives.

In response to the changing fiscal environment, cities on average are making greater use of public pricing and privatization. But in the long run, it is not likely that the exogeneous shocks can be overcome solely by changes in financing schemes. We predict a reduction in the level of public service delivery will be necessary in order to balance local budgets. In the final section of this chapter, we speculate on possible consequences of this radical change in local public service delivery.

EXOGENEOUS FORCES
AFFECTING PUBLIC SERVICE DELIVERY
IN CITIES

There are myriad factors that affect how adequately local public officials provide services. Some of these factors are well within the control of public managers, but others are beyond their control. Four major exogenous factors are (1) the energy crisis, (2) inflation, (3) the taxpayer's revolt, and (4) demographic shifts in popuation. Of course, these factors will always cause some problems in service provision because of a shortage here or there, taxpayer's complaining, or people moving in or out of cities. However, the intensity of these four exogeneous factors has been particularly forceful in recent years and prospects are not promising for a reduction in this intensity.

Resource scarcity appears to be at the heart of the problem. Today it takes far more economic resources than in the past to recover a diminishing amount of nonrenewable energy. Until feasible lower-cost substitutes are developed and marketed, all sectors of the economy that depend on energy inputs will confront the growing problem of resource scarcity.

Within this backdrop of growing natural resource scarcity governments—federal, state, and local—also will be coping with short run increases in demand for public services. Only some of these demands will be met, as the levels of government attempt to sort out and define their separate roles under a new awareness of scarcity. The difficulty of this undertaking will be compounded because the experience of governments over the last two decades has been one of relative resource plenty, and this has caused public expectations to rise. While other sectors of the economy stagnated or grew only slightly, the public sector experienced a relatively high growth rate. Between 1960 and 1980, state and local government expenditures increased from 9.2% of the GNP to 12.8 %. State and local government employment more than doubled, increasing 118.7% (Council of Economic Advisors [CEA], 1981). But this increase in employment under inflationary times is a mixed blessing. Some have argued that the public sector is hurt more than the private sector by inflation because the local public sector is so labor intensive (Cohen and Levinson, 1980).

Another reason why the local public sector may be disproportionately hurt by inflation is that resource scarcity has prompted many states to limit local government growth. The taxpayer's revolt movement has been successful in getting taxing and spending limitations on many state ballots (Mikesell, 1979). However, constraining local governments to no-growth in inflationary times results in real declines as a fixed sum of money purchases less and less.

Some have argued that the taxpayer's revolt movement represents a growing frustration with too much government intervention at all levels. California's Proposition 13 has been viewed as more of a reaction against the federal government's intervention into people's lives than a reaction to reduce local government provision of traditional public services (Mushkin, 1979: 154-55). Voters, not able to change federal government policies, could show their displeasure by a vote against a government closer to home.

Frustration with reduced economic opportunities, caused partially by scarce resources, has prompted many people to move from cold northern cities to sunny southern cities. These demographic shifts will place even more demands on local public sectors in the South while northern cities, suffering a reduced tax base, must still provide services to remaining residents. Thus the general problem of resource scarcity manifested in the energy crisis, inflation, the taxpayer's revolt, and demographic shifts will play an important role in reshaping how well cities provide services in the twenty-first century.

Figure 18.1 represents a framework for viewing the factors that affect local public service delivery. The four exogenous factors and the federal government's effort to redefine federalism will change the pool of available resources. This pool will be reduced by some initiatives and expanded by others. Still other political efforts, not directly related to the redifinition of federalism, like reduction in federal regulations, will also change the pool. The net effect, however, will be a smaller pool of resources. Local governments will react to

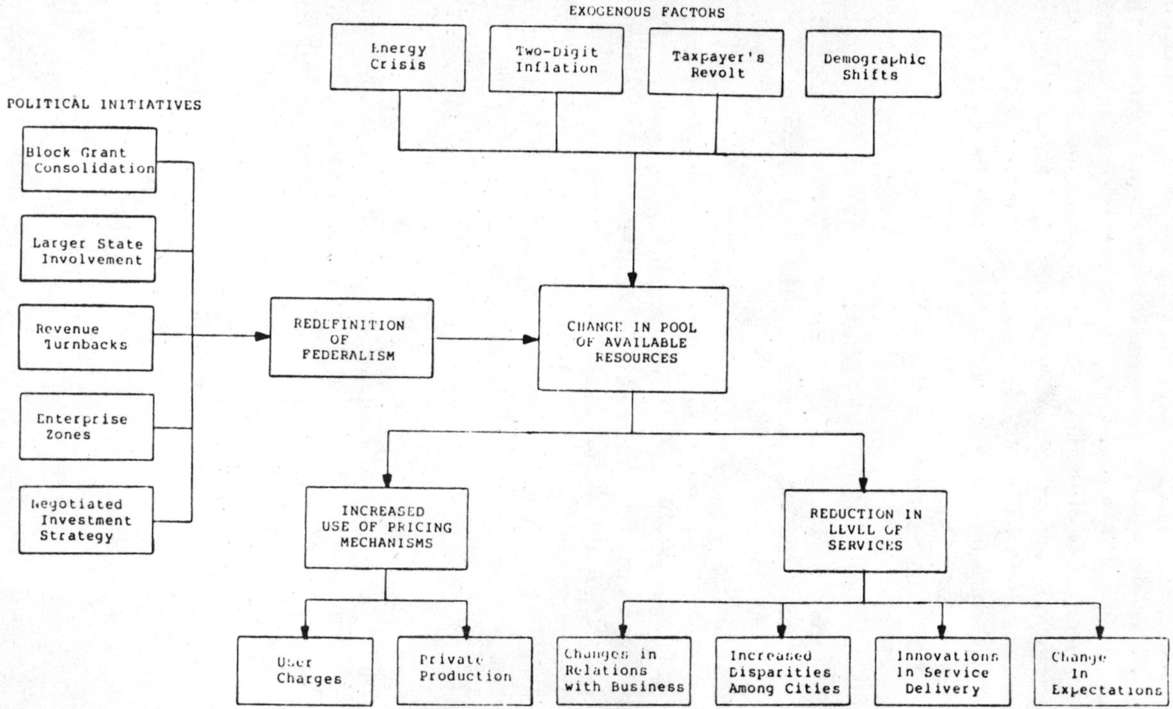

FIGURE 18.1   Cities' Resources for Public Services

declining resources by using nontraditional means of financing and service delivery. But this effort will not be strong enough to offset the full reduction in resources, so service levels will be reduced. This decline in services will be a mixed blessing that increases disparities among cities, yet creates an environment for greater innovation.

THE EVOLUTION OF THE FEDERAL ROLE IN CITIES

A major factor affecting local public service delivery is the current redefinition of the federal role in intergovernmental relations. This redefinition has proven to be a catalyst in hastening a change in the way local services are provided by changing the way these services are financed. The disturbance caused by changes in the federal-local role is very real and may well mark the beginning of a revolution in local public service delivery.

In order to place this change in proper perspective, it is useful to discuss how the federal government has evolved from a distant observer of local affairs to a force directly involved in local governance. In particular, it is useful to understand how the federal government's activities have become "more pervasive, more intrusive, more unmanageable, more ineffective, more costly, and above all, more unaccountable" (Advisory Committee on Intergovernmental Relations [ACIR], 1980a: 1).

Federal involvement in state and local affairs has not always been intrusive. Intergovernmental relations prior to the constitution were characterized by a relatively weak federal government. During the Colonial period a strong pattern of local self-government emerged, and this pattern was not dissolved when the Revolutionary War was won. The drafters of the Articles of Confederation, fearful of national rule, made sure that the Continental Congress was not a strong one. But the division of responsibilities between the federal and state governments was not a functional one, and in 1789 a new governmental system was adopted that reserved some powers to both the federal and state governments, and called for a sharing of others. Cities, not officially recognized in the constitution, were established as creatures of the states.

Two major events that had a profound effect on the role of the federal government were the Civil War and the Great Depression. The Civil War cleared the way for a federal assumption of greatly broadened fiscal responsibilities (Break, 1980). The Morrill Act of

1862 established land grant universities, and in the second half of the century, the federal government began to take an active role in national economic development of transportation infrastructure, such as railroads and canals, inland waterways, and, eventually, highways.

The Great Depression and the consequent New Deal marked another significant expansion of the federal role in intergovernmental relations. The federal government began to use aid for relief, reform, and economic recovery. Usually this aid went to the states in the form of categorical grants, ear-marked for specific purposes with strings attached. Very few grants went directly to localities, and those that did dealt only with urban renewal and public housing (Judd and Kopel, 1978: 168). Prior to 1960 only fifteen or so programs of local aid involved no state role at all (Walker, 1981: 82). These included the Housing Act of 1937 and 1949, the 1946 airport legislation, and the urban planning and renewal program of 1954.

The era from 1960 until the late 1970s was characterized by increasing intrusiveness of the federal government into all aspects of the federal government into all aspects of the public sector. Congress became the center of the intergovernmental system. Federal involvement with local governments blossomed, surpassing the states' role. For the first time states were entirely bypassed when localities dealt directly with federal bureaucrats to receive certain federal funds. Federal aid, however, was not a free gift to localities. Cross-cutting requirements were established that required recipients to adhere to a variety of national objectives generally irrelevant to the espoused purposes of the grant, such as nondiscrimination and environmental protection (Peirce and Hamilton, 1981).

Federal grants to localities accounted for their largest single source of revenues, and some, if not many, cities were hooked on federal aid (Shannon and Ross, 1976). The swing was now complete; the federal government had evolved from being a weak link in the intergovernmental system to being a prime mover in state and local affairs. But factors exogenous to the federal government developed that made it difficult to continue ever-increasing funding of state and local activities. By 1979, the real size of federal outlays to states and localities began to decline. The taxpayer's revolt had sent a message to Washington: The public sector was perceived as being too large, crowding out the private sector. Politicians, sensitive to this message, began to espouse the virtues of public sector frugality, and the pendulum came to a crest and began to swing back.

INITIATIVES FOR
A REDEFINITION OF
THE FEDERAL ROLE

Initiatives are emerging now that could reshape the economic and political climate for many years to come. We think there are at least five that bear close watching because of their potential effects on how cities will function in the future vis-a-vis higher levels of government. These are (1) the movement towards block grant consolidation, (2) the potential for larger state involvement in the financing and goverance of cities, (3) revenue turnbacks, (4) enterprise zones, and (5) negotiated investment strategies.

BLOCK GRANTS

The Budget Reconciliation Act of 1981 consolidated a number of categorical grant programs into block grants. Although the Reagan administration proposed a wider consolidation than Congress granted, the approved FY82 budget marks a monumental step in the effort to reduce federal involvement. Instead of five block grants consolidating 88 programs, Congress approved the consolidation of 57 programs under nine block grants. Some of the largest categorical programs of the initial 88 remained categorical (Stanfield, 1981a).

Federal involvement has been reduced in at least two ways. First, the dollar value of federal support was decreased, both in nominal and real terms; second, and perhaps even more significant, grantees' administrative requirements were lessened, although not by as much as the administration proposed. Both reductions mark a reversal in the trend of more and deeper federal involvement, and both reductions, assuming the new trend continues for some time, will greatly affect the way cities provide services in the next century.

In the future fewer strings will be attached to federal involvement, thus giving cities a greater freedom of choice in the ways they will respond to the demands of local residents. Cities, however, may not have a completely free reign because the authority vacuum created with reduced federal involvement may be partially filled by increased state involvement.

ENCOURAGEMENT OF LARGER STATE INVOLVEMENT

States have the potential to assist cities in adjusting to the significant decreases in federal aid in at least two important ways. First, states have the power to reform local fiscal systems or grant broader local authority enabling cities to tackle their fiscal, structural, and functional problems. This is a structural issue that state legislatures must consider. Second, state aid to localities has much potential for providing fiscal relief to financially hard-pressed cities. Whether or not the states choose to extend this type of aid is a matter of both capacity and the willingness to take on this new role.

States must have both administrative and fiscal capacity to provide relief to fiscally distressed cities. The argument over state's capacity to take on this new responsibility is an old and impassioned one, and there is much variation between the states. Some analysts feel that state governments need to undergo substantial political and philosophical reorientation before they can assume a more significant role in federalism (DeGrove and Stroud, 1981). Others feel it may not be a matter of administrative ability, but rather, of fiscal resources (Stanfield, 1981b: 830). Falling revenues are combining with an inability to borrow in a way that makes it extremely difficult for states to respond.

Also at issue is the states' willingness to assume this responsibility. Many argue that it was state neglect of local governments in the first place that caused the federal government to step in (Stanfield, 1981b: 831). Others contend that while this may have been true in the past, significant changes have occurred in state governments (Walker, 1980). There is a growing record of some innovative state action which demonstrates their capacity and willingness to respond to local problems (Warren, 1980, 1981a).

REVENUE TURNBACKS

The second phase in the movement to reduce federal involvement is to turn back revenues to state and local governments. There are a number of turnback methods, each serving different policy objectives (Davis and Shannon, 1981). Revenue sharing, a program in existence since 1972, returns revenue through a series of distribution formulas designed to equalize on the basis of need and fiscal effort (Noto, 1981). Another method is tax sharing, whereby the federal government simply returns a share of federal taxes to their geographic origin. Under this method no attempt is made to correct for horizontal imbalances in the delivery of local public services or the possibility that all taxes received from a geographic area are not ultimately borne by residents of that area. Relinquishment of a federal tax is a third turnback method. Here the federal government could give up all or part of a tax, requiring the state or locality to pick it up as an additional revenue source. Or, the pick-up requirement could be ignored, resulting in a direct turnback to taxpayers if the state or locality chose not to pick it up.

The idea of revenue turnback is not completely new. The concept is currently in use in Canada, where the federal government has yielded percentage points of the federal personal and corporate income taxes to the provinces. It also makes equalizing payments to provinces with tax yields below some standard (ACIR, 1981: 39-56). In this country the Eisenhower administration had raised the issue, but was unsuccessful in getting its program accepted (Stanfield, 1981c). In the 1950s, however, the functions for each level of government were less commingled. Now the emphasis is on sorting out and redefining the functions, along with freeing up resources for each level to carry out its specific functions. Thus revenue turnbacks may become an important source of revenue for cities in the future.

Although revenue sources for cities in the twenty-first century will not be radically different from today's revenue sources, the shares that these sources produce for the total budget may be quite different. One way that the shares could change is for cities to reduce reliance on a particular revenue source from a specific geographic area, thus promoting economic development by creating a zone where residents and businesses have a reduced public sector burden.

ENTERPRISE ZONES

In order to lure the private sector into urban areas in need of development, the appropriate incentives must be created to offset current market incentives that dictate use of more profitable urban locations. One British idea, which demands only indirect federal involvement, is to free certain designated urban areas from taxes and government regulations (Breckenridge, 1981). This would establish an enterprise zone where the private sector could invest and create jobs without being encumbered by undue government oversight. Advocates of this controversial concept believe a reduction in government intervention will provide the appropriate incentives to make distressed urban areas prime candidates for private redevelopment. Skeptics point out that a hidden subsidy of unknown budgetary impact is being given. They claim that a grant program, like the Urban Development Action Grant program (UDAG) established under the Carter administration, is a more appropriate subsidy. The reason is that both recipient and amount are known before the grant is approved.

Even though Congress has not enacted into law a federal enterprise zone bill, some states and cities are already planning their own versions of enterprise zone. The extent to which this concept could help to reshape the face of cities in the twenty-first century is still uncertain, but even in the conceptual stage it has compelled government and business to sit down and discuss what types of incentives would most likely achieve the desired goals. Another idea that would bring together the three levels of government and the private sector is the concept of a negotiated investment strategy.

NEGOTIATED INVESTMENT STRATEGY

Local government agendas are often sidetracked because all the actors affected by the agenda were not involved in the process of agenda setting. A resolution of potential conflicts before an agenda is set would lead to far more efficient delivery of local public services. The concept of negotiated investment strategy goes further than previous attempts to lever the private sector. It is an innovative approach to involve all the actors in a mediated process that attempts to arrive at an agreed course of action.

As the federal role in city government changes from actively setting agendas to one of relative noninterference, locally determined demands stand a better chance of being met if a consensus is reached on the local agenda. The negotiated process could become the keystone in establishing an often stated objective—a true "public-private partnership" (Warren, 1981b).

REDUCTION IN THE LEVEL
OF PUBLIC SERVICE DELIVERY
IN CITIES

We have described factors, both economic and political, that are outside the control of city officials. Officials, faced with a net decrease in the pool of available resources, are responding to this situation in two ways: trying to expand revenues, and trying to decrease spending. First, city governments are exploring various financing mechanisms that they can implement without coordination with other levels of government. These include public pricing and privatization. Between 1955 and 1980 city government reliance on user charges grew from $0.14 per dollar of tax revenue to $0.32 per dollar of tax revenue, a growth of 113%. Prospects are promising for a continued expansion in the future (Mushkin, et al., 1979). Cities also have turned over various functions to the private sector, hoping that the services can be provided more efficiently. Cities in the future will have to rely more on these direct-benefit means of financing, if they are to minimize service level reductions.

With so many upheavals occurring at once it is difficult to speculate on the scope of city government in the twenty-first century. There are, however, four consequences that seem likely to occur: (1) changes in private-public relations, (2) increases in disparities among cities, (3) innovations in service delivery, and (4) changes in expectations.

One of the consequences of the reduction of local public service levels is that there will be a changing relationship between cities and their business communities. Businesses are dependent upon cities for the provision of public infrastructure and services. Providing adequate infrastructure is an important way that cities can contribute to reindustrialization. To the extent that local governments will not be able to provide these services, business will either have to become more involved in the community and aid in the financing or production of services, or move to another community that is able to provide the desired services.

The likely response of businesses will depend on the severity of the fiscal distress of the municipality, the mobility of the firm, the cost of transferring operations to a locality that provides the desired bundle of services, and the commitment of the firm to the community in which it is located.

Another consequence of reduced levels of public service delivery in cities will be increasing disparities between cities. The revenue short-fall is not likely to affect all cities in the same way and to the same degree. Those cities that traditionally provided a relatively high level of services, like many cities in the Northeast and Midwest, will be particularly hard hit.

A third consequence of reductions in public service delivery will be innovations in the way services are delivered. These could range from the private production of services traditionally supplied by the public sector, as previously mentioned, to the regionalization of some ser-vices by neighboring communities who wish to eliminate duplication of effort, to completely new methods of provision. Successful in-novations will bring more efficiency to the management of the local public sector (Pascal, 1980).

A fourth consequence of the reductions in public service activity, and potentially the saving grace of the local public sector, will be a lowering of citizens' expectations as to the appropriate scope of ac-tivities to be performed by the public sector. In the past, public demand was a prime force leading to the expansion of the local public sector. When, out of necessity, these demands are attenuated, the local public sector may finally be able to adjust to a more stable, lower level of activity.

In conclusion, all of these likely consequences focus on the crucial issue of redistribution—redistributing scarce resources among the various sectors of the economy and redistributing responsibilities among the various sectors of the economy and redistributing responsibilities among the various levels of government. The future composition of city public services will depend, to a large degree, on how the reduced level of resources is redistributed. Some have argued that state and local governments have already suffered disproportionate reductions, especially in the area of caring for the poor (Farber, 1981). Others have argued that resource constraints can help to create a more effective public sector, provided new patterns of intergovernmental and interinstitutional relationships are found (Gappert, 1979). Still others feel that the key to the viability of our urban centers is the ability of our nation's industries to keep up with world and domestic opportunities, and that the federal government should focus revitalization efforts not on areas, but rather, on industries and sectors of the economy (Schwartz and Choate, 1980: 76-82). The net effect of all of these factors on any particular city's ability to deliver public services in the future will also depend on the city's administrative and management talent and the political and fiscal constraints within which the public servants must operate.

REFERENCES

Advisory Commission on Intergovernmental Relations (1980a) The Federal Role in the Federal System: the Dynamics of Growth. Washington DC: Government Printing Office.

—— (1980b) Significant Features of Fiscal Federalism, 1979-1980. Washington DC: Government Printing Office.

—— (1981) Studies in Comparative Federalism: Canada. Washington DC: Government Printing Office.

BLAIR, J. P. and D. NACHMIAS (1979) "Urban policy in the lean society," pp. 11-42 in J. P. Blair and D. Nachmias (eds.) Fiscal Retrenchment and Urban Policy. Beverly Hills, CA: Sage.

BREAK, G. F. (1981) "Fiscal federalism in the United States: the first 200 years, evolution and outlook," pp. 39-66 in Advisory Commission on Intergovernmental Relations (ed.) The Future of Federalism in the 1980s. Washington DC: Government Printing Office.

BRECKENRIDGE, C. (1981) "Enterprise zones as a concept." Congressional Research Service (March).

BRENNAN, G. and J. BUCHANAN (1979) "The logic of tax limits: alternative constitutional constraints on the power to tax." National Tax Journal 32 (June) (Supplement): 11-22.

Business Week (1981) "State and local governments in trouble." October 24: 135-181.

COHEN, H. A. and R. D. LEVINSON (1980) "The impact and interaction of inflation with state and local funding systems." National Tax Journal 33 (September): 279-289.

Council of Economic Advisors (1981) Economic Report to the President. Washington DC: Government Printing Office.

DAVIS, A. J. and J. SHANNON (1981) "Stage two: revenue turnbacks." Intergovernmental Perspective 7 (Spring): 18-25.

DEGROVE, J. and N. E. STROUD (1981) "Local and regional goverance in the changing federalism of the 1980s," pp. 45-59 in American Federalism in the 1980s: Causes and Consequences. Cambridge MA: Lincoln Institute of Land Policy.

FARBER, S. B. (1981) "Reforming and cutting back the federal aid system: implications for state and local governments." National Tax Journal 34 (September): 311-314.

GAPPERT, G. M. (1979) "Employment policy in postaffluent America," pp. 159-182 in J. P. Blair and D. Nachmias (eds.) Fiscal Retrenchment and Urban Policy. Beverly Hills, CA: Sage.

JUDD, D. R. and F. N. KOPEL (1978) "The search for national urban policy: from Kennedy to Carter," pp. 163-200 in T. J. Lowi and A. Stone (eds.) Nationalizing Government: Public Policies in America. Beverly Hills, CA: Sage.

MIKESELL, J. L. (1979) "The season of tax revolt," pp. 107-130 in J. P. Blair and D. Nachmias (eds.) Fiscal Retrenchment and Urban Policy. Beverly Hills, CA: Sage.

MUSHKIN, S. J. [ed.] (1979) Proposition 13 and its Consequences for Public Management. Washington DC: Council for Applied Social Research.

—— F. H. SANDIFER, C. L. VEHORN, and C. G. TURNER (1979) "The taxpayer revolt: an opportunity to make positive changes in local government," pp. 119-140 in S. J. Mushkin (ed.) Proposition 13 and its Consequences for Public Management. Washington DC: Council for Applied Social Research.

NOTO, N. A. (1981) "Revenue turnbacks: an evaluation according to traditional criteria for intergovernmental funding." Congressional Research Service, October 19.

PASCAL, A. H. (1980) "User charges, contracting out, and privatization in an era of fiscal retrenchment." Santa Monica, CA: Rand.

PEIRCE, N. R. and J. HAMILTON (1981) " 'Flypaper federalism'—states, cities want to shed rules that accompany aid." National Journal (September 12): 1636-1639.

SCHWARTZ, G. G. and P. CHOATE (1980) Being Number One: Rebuilding the U.S. Economy. Lexington MA: D. C. Heath.

SHANNON, J. and J. ROSS (1976) "Cities: their increasing dependency on state and federal aid," pp. 189-212 in H. J. Bryce (ed.) Small Cities in Transition: The Dynamics of Growth and Decline. Cambridge, MA: Ballinger.

STANFIELD, R. (1981a) "For the states, it's time to put up or shut up on federal block grants." National Journal (October 10): 1800-1805.

—— (1981b) "Block grants look fine to states; it's the money that's the problem." National Journal (May 9): 828-832.

—— (1981c) "Ready for 'New Federalism,' phase II? Turning tax sources back to the states." National Journal (August 22): 1492-1497.

U.S. Department of Housing and Urban Development (1980) Urban Development Action Grant Program: Second Annual Report. Washington, DC: Government Printing Office.

WALKER, D. B. (1980) "The states and the system: changes and choices." Intergovernmental Perspective 6 (Fall): 7-12.

—— (1981) Toward a Functioning Federalism. Cambridge, MA: Winthrop.

WARREN, C. G. (1980) "The states and urban strategies: a comparative analysis." Washington DC: U.S. Department of Housing and Urban Development, Office of Policy Development and Research.

———— (1981a) "The changing federalism of the eighties: the future of state governments," p. 27-42 in American Federalism in the 1980s: Causes and Consequences. Cambridge, MA: Lincoln Institute of Land Policy.

———— (1981b) "National implications of a negotiated approach to federalism." Presented at "Mediating Federalism: Roundtable on Negotiated Investment Strategy," sponsored by the Charles F. Kettering Foundation and the Academy for Contemporary Problems, July 8.

19

Epilogue:
Urban Management and the Future

GARY GAPPERT
RICHARD V. KNIGHT

☐ HISTORICALLY the field of urban management has been primarily concerned with the management of growth. In the next two decades, although growth will be continuing in portions of our national system of cities, the primary concerns for urban management will be with either (1) the management of shrinkage, and/or (2) the management of the transformation of the city to a different set of relationships in the evolving national and transnational system of cities.

In the last decade the private sector has responded to the new economic turbulence by developing strategic planning systems. Strategic planning and strategic management require new organizational frameworks and values. Transformation management in the public sector is implicitly underway but is obscured by the conflict and crisis-management concerns generated by short-term fiscal instability and uncertainty. The paradox of strategic planning is that it is most required when the conditions to do it are frequently the worst (Gluck, 1982).

American urban development will require a fundamental reorientation similar to that experienced by corporations that have embraced strategic management. The successful city of the future will have to be an inttentional city. It will require new concepts, perhaps a new language, and a new political will. By the year 2000 new types of urban structures, functions, and institutions will have been developed by many communities.

In this epilogue three issues are raised. First, the requirements for a new discipline of city development and design are discussed.

Second, some of the emerging skills for a new style of urban planning management are identified. Finally, an urban futures management model is outlined. The perspective of this chapter is illustrated in Figure 19.1. A new, emerging philosophy of urban development, translated by the political conditions of the post-affluent transition creates new demands on urban managers and planners. Depending upon the vitality of urban institutions and their level of competence, some cities can be expected to deal with the complexity of designing a realistic future.

CITY DEVELOPMENT AND DESIGN: TOWARD A NEW DISCIPLINE

A new, broadly focused philosophical orientation is needed in order to advance the state of the art of city development and design. One of the main reasons why thinking about cities is so difficult is that the discipline needed by those engaged in designing and developing an individual city is in its formative stage. The intellectual rigor and empirical and quantitative orientation of economics is necessary but not sufficient. Political economy has indeed yielded important insights into the processes of city development, but none of the more established disciplines can encompass a subject as broad and multifaceted as city development (Knight, 1979).

Although critical attributes of this new discipline are becoming clear, synthesis still lies ahead of us. But at least seven attributes can be discussed. First, city development and design by its nature requires a holistic approach. A disciplined approach to cities must be holistic because it requires the synthesis of knowledge gained from a broad array of traditional disciplines in arts, sciences, engineering, business, and professional colleges. Clearly, the synthesis of knowledge and the application of this knowledge to city issues is in itself a new discipline. Whereas the integrity of the traditional departmentalized disciplines, such as economics, political science, sociology, architecture, planning, and history, dictate an analytical approach, that is, the breaking down of phenomena into small and more manageable components, the integrity of city development dictates more of a generalistic or synthetic approach. City development seeks to understaand the multifaceted and pluralistic nature of evolving human settlements and changes that occur between the many interdependent components of which they are comprised. The city's problems cannot be resolved in isolation of each other or of the general conditions that produce or contribute to them. The holistic approach considers the many diverse

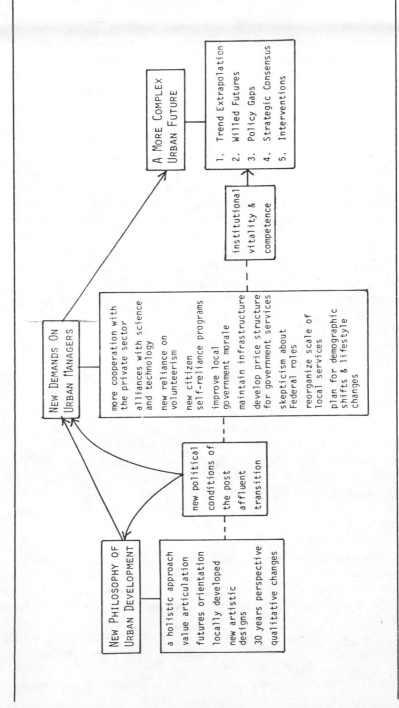

FIGURE 19.1 A Perspective on Urban Management
SOURCE: Richard V. Knight, *The City and Global Economy*, forthcoming.

facets of the city, including its values and concerns, and provides a framework for forging conflicting interests into common goals and synthesizing them into a sense of community interests.

Second, one of the key aspects of city development is the identification and articulation of the values on which the city is founded. This is not a simple process; it requires considerable skill and sensitivity to identify values that underlie a city's strengths and weaknesses. If, for example, certain values underlying development patterns give rise to certain problems, then the first step in the long process of correcting the situation is clearly to identify those values and document their effect on development patterns. As awareness of the consequences and social costs of old values becomes more widespread in the community, they may gradually give way to new values that do not have the same consequences. Careful conceptualization and documentation of conditions that generate problems will eventually lead to the modification of the values and concomitant processes that generated them. Problems once understood will not have to be repeated in future plans; they can be designed out.

Third, city development and design is both process and future oriented. This is very different from most traditional disciplines, which tend to emphasize comparative studies, tend to be past oriented, and tend to concern theoretical rather than applied research. Moreover, they are not usually concerned with changes in or the evolution of values, a key element of any future-oriented studies. Economics, for example, is becoming increasingly dominated by methodology. Elaborate econometric models do not account for institutional or technological changes that tend to reshape market forces and change the industrial and spatial organization of the economy. Cities are open economies and are differentially affected by changes in social values, prices, behavior, government regulations, and other factors that affect the allocation of resources. Moreover, if the city is viewed as a cultural entity, as the anvil of a civilization, as the place where values are forged as, then differences in values among cities become key to understanding the nature of their development.

Unless a city appreciates its heritage, its contributions to society, its values; unless a city understands the nature and courses of social pathologies that have arisen from past values or growth practices; unless it can formulate a meaningful and widely shared vision of its future, it will not be able to discipline its future development or control its own destiny. The city that has not examined or cannot effectively articulare the values that drive or determine its development, will not be able to design its future or defend itself against powerfull outside forces. Such a city will, in effect, be adrift in a turbulent environment, reacting to forces that are not understood. It will be open to initiatives of outsiders and unable to prevent others from imposing their values and solutions.

Fourth, not only does the discipline need to be future and value oriented, but it must be established locally so that local values, local development processes, and local concerns will be reflected. Values, development processes, and community needs differ significantly by city and change over time as the city evolves. In order to identify the document these differences and monitor changes over time, a locally based client or sponsor is necessary so an established intelligence base can be erected and continuously updated over an extended time period. A problem that arises, however, is that the composition of the client changes as the city evolves; the old client or sponsors may not be interested in investing in an undertaking which, if successful, will result in change. This problem is particularly acute in the older, heavily industrialized areas.

Initiatives taken on behalf of a city or of groups of individuals within a city need to be of local origin, cognizant of development processes at play, and closely attuned and oriented to the city's present values and futuree needs. But emerging cities are often apprehensive, insecure, and try to avoid parochialism by turning to nonresident national experts. As a result, many cities emulate the style, strategies, and values of other cities even though such approaches do not relect the strengths of an indigenous nature. Investments in seeking outside solutions to local problems tend to divert local funds from basic research or monitoring of the development processes. Unless the client city is able to articulate its own goals and values and clearly define and document its environment, outside professionals will have difficulty in contributing their particular expertise.

The problem with outside studies is that the knowledge developed leaves with the consultant while the study gathers dust. The idea of importing experts and creating instant think tanks is attractive, especially in times of crisis, but such policies are not likely to lead to solutions that are attuned to the peculiarly local character of the problem and may overlook the possibility of new indigenous approaches to city development. In many cities, attempts at implementing national solutions to localized problems have contributed to the frustration and disillusionment of those involved.

A fifth attribute of the discipline of city development and design that is particularly important in the industrial city is that entirely new types or species of world cities are being formed. These cities are still undefined.

Artists play a key role in this aspect of city development, but their contribution has yet to be sought. Compared to the millions that some cities are spending on promotion, very little is invested in supporting those who could best articulate positive values of the industrial order or of the culture of their particular cities. A sense of identity that inspires civic pride has, in the past, been created by artists, writers, architects and philosophers. One of the great challenges of this generation is to conserve the legacy of the industrial revolution and to restore citizens' pride, confidence, and trust in their troubled industrial cities. We may have wondered how other cities such as London, Paris, Rome, Florence, Athens, or Canton project such positive images. They also went through development crises in the past. But that was before jingles, bumper stickers, T-shirts, and other hype approaches were in fashion. Emphasis on such costly programs must be questioned because they do not have a lasting effect as do the works of a Da Vinci, Michelangelo, Dickens, Wren, Balzac, Sullivan, Burnham, Olmstead, or Frank Lloyd Wright. The work of artists endures because it gives form to the new values of their time. Clearly, artists could play a greater role in improving the content of communications programs that are implemented by advertising and public relations experts. The values on which the industrial order and cities are founded have to be articulated before they can be communicated. Successful marketing of a new type of city depends in large part on how well the city has been defined.

The sixth attribute of city development and design concerns the time perspective required. Time is particularly important when one starts to think seriously about designing the industrial city of the future. Planning in most American cities is a function of the priorities of key elected officials. Since most mayors face reelection every two or four years, this restricts the kind of projects or plans that can be seriously considered and those that can be completed before the next election campaign. The appropriate time for city development and design in older industrial cities is, however, more like thirty or fifty years—ten or twenty times the present political time frame.

It is very instructive to inquire how cities that are so attractive today were able to design and transform their environments. Most went through a long, difficult restructuring process. In those cities that had great fires such as Rome, London, Chicago, Toronto, and San Francisco, in those which were paralyzed by civil chaos, as in Paris, and rebuilt by dictators placed in power to restore order, and in those devastated by war and rebuilt by postwar planners, the time frame may have been somewhat reduced. But in a democratic society where market forces dominate, the restructuring of an advanced industrial region is necessarily an incremental process. In such situations long-range planning is critical to continued development.

Of all the lessons that can be gained from the experience of older, more mature cities, the need to extend the present time frame is probably the most important. The planning horizon determines in large part what can be accomplished. There is a Chinese proverb that says, "If you are planning for one year, plant grain; if you are planning for ten, plant trees; and if you are planning for a hundred years, plant men." The type of institutional building and upgrading involved in developing the industrial city requires long-term commitments by individuals who can implement long-term programs.

A seventh attribute of the disciplines that would facilitate our thinking about cities is that as we shift from a growth-oriented society to one that is concerned more with development, we have to change our mode of thought from one concerned with quantitative change (increasing scale) to one concerned with qualitative change (changing behavior). Rather than being concerned with problems associated with the assimilation of new immigrants and the expansion of the physical and social infrastructure of a region, we now have to focus more on ways to upgrade organizations and to improve the quality of life of those that are now living in the region. Industrial central cities are, in fact, becoming smaller in terms of number of people as they develop, but the composition of their population is changing from working to middleclass. The old infrastructure and old political arrangements may not be appropriate in the new industrial city. Many urbanologists such as Eric Lampard, Edgar Dunn, Russel Ackoff, and Edward Banfield argue that development consists primarily of a learning process. New associations, new institutions, and new settings have to be created.

Louis Mumford, one of the most renowned experts on city development, noted that the tendency to planless dispersion is worldwide, and that "in a hundred futilre ways people seek an individual solution to their social problems and so ultimately create a second social problem." How can the difficult decisions concerning the design of the industrial city be made so that social and individual benefits are maximized, while political externality costs are minimized? What endears the city to later generations is its power to master its own destiny and to express its best ideals in the transformation of its environment.

These observations concerning the nature and need for forming a new discipline, city development and design, will indicate the nature of the commitment that will have to be made if American cities are to realize fully their potential as world-class cities.

URBAN MANAGEMENT: NEW SKILLS IN OLD PLANNERS

While it may be difficult to achieve a new philosophical orientation to the problems, and opportunities, of urban development, there are some emerging trends that represent significant issues for the management of urban institutions in the immediate future.

First, the intense and rapid growth in state and local government is likely to level off. There are now about 12-13 million full and parttime workers in state and local government, while the federal civilian labor force has remained stable at about 2.8 million since the mid 1960s (Gappert, 1980).

The new demands of the New Federalism, combined with cutbacks in federal funds and the discretionary nature of block grant funding, means that there are likely to be severe pressures to change the composition of the state and local governmental work force to bring it into alignment with changing functions and new priorities.

Second, it is likely that municipal administration will increasingly require interagency cooperation. Different types of management teams, both formal and informal, will be required.

Along with the above condition, different budget processes will be developed. There should be a renewed interest in program and performance budgeting. With the new scarcity of resources, demands to monitor the actual work of public employees may increase.

It is also likely that the distinctions between suburb and city will be less sharp. Increasingly, local governments in the context of a megalopolis or a metroplex will find a need for regional problem solving of common problems. The inner suburbs, in particular, will come to share more problems with central cities. Exurban enclaves will be "adopted" by corporate sponsors, who will concentrate specialized industries in massive campus-like complexes away from the congestion and development controls of the older suburban rings. The Monsanto complex outside St. Louis is one example, as is the Johns Mansville Facility outside Denver.

The political climate will also begin to reflect more substantial intergenerational conflicts. The elderly and retired persons are finding new ways to influence and participate in local government. New problems are emerging among younger teenagers. Adults without children are a larger proportion of the local electorate. Generational differences in lifestyle, values, and service requirements will become a more pronounced force in local politics.

Issues of personal conduct are also likely to become topics of local governmental debates. Currently they cover broad areas, ranging from the common use of leisure space and time to new constraints on pet owners in New York City. The designation of nonsmoking areas is another serious issue in some areas. Health and child care are also causing concern. The epidemics of teenage pregnancies and drug and alcohol abuse, for example, are creating many local problems that will become areas of public debate. It is unclear what tensions and solutions may result from these various concerns, but communities may increasingly seek to establish codes of behavior in areas in which there is perceived to be an overriding public interest. Proposals are heard for an urban code of humanistic conduct.

Associated with the concerns discussed above, there will be new efforts to manage and affect citizen participation. In the last decade many new approaches to community planning of particular services have been initiated because of federal and state mandates. The total effect of this participation is not clear but, in any case, some participation, especially at the neighborhood and block level, will continue. And as the housing shortage keeps more people into the large apartment complexes that ring the central city, new forms of middle-class organization and participation are likely to emerge.

Women will become much more active and involved in local governance and management. It should be no surprise if by the year 2000 half of the governors and mayors are women. A similar projection could be made for the urban management profession.

Another likely shift in participation can also be identified. This relates to the growing credibility of celebrities and athletes as political candidates. A certain kind of irony exists when public credibility in government is restored when a reassuring figure from the video media is elected to office, endorses a candidate, or promotes an issue.

Another need will be that of orchestrating coalitions among citizens with sharply differing values and lifestyle. With some expected decrease in mobility, communities will continue to need to develop a toleration for diversity and social deviation. The ongoing problems and crises of the post-affluent transition, together with their effects on economic and environmental well-being, will require that public managers have the skills to articulate and interpret the needs of their communities.

These needs will be identified as planners address the specific future realities of their communities. Some aspects of the future are likely to be unpleasant. New forms of corruption and new forms of patronage may be forthcoming. New types of community fanaticism may occur. Enclaves of radically eccentric groups may not be unusual in large metropolitan areas.

Every major trend seems to point to a greater degree of municipal governance in the future. There will also be a greater number of sources of input and fewer areas of autonomy in the management of large urban communities. Professional rationality and performance may have to be redefined in order to deal effectively with the new conditions.

Rutter (1980) has indicated that the next several decades are "a time for vision, imagination, and creativity." As he writes:

> The leadership of the future will have many dimensions, as it does today, and as it did yesterday. A portion will be political, exercised by elected officials. A portion will be popular, assumed by involved and informed citizens. A portion will be technical, provided by experts in service delivery. And a portion will be managerial, cutting across the political, popular, and technical.

Rutter also reports that "the promise of the future for any person in a leadership role is a life of increased stress." He proposes an interesting metaphor—the need to "nuture" the essential community. In his analysis urban leaders must learn to manage without illusions, and to get by modestly, by striving for:

—more cooperation with the private sector,

—stronger alliances with the scientific and academic communities,

—greater emphasis on volunteerism,

—helping citizens do for themselves what they have come to expect from local government,

—reexamining public policies that implicitly assume a no-risk base,

—making local government employment more satisfying,

—maintaining the infrastructure,

—regulating the demand for government services,

—being skeptical about central government support,

—reexamining the whole scale of local government,

—altering services in anticipation of the demographic shifts over the next twenty years.

These new facets of urban leadership are evolving incrementally. What is also management and urban incremental in its nature is the emergence of a new vision of the city and its appropriate role in the society of the twenty-first century. The new discipline of urban development and the emerging skills of a new breed of urban manager will only be truely effective if they are supportive by a new strategic consensus about urban functions, structures, and, most importantly, institutions.

AN URBAN FUTURES MANAGEMENT MODEL

The perspectives of futures research can be applied to urban management and development in at least three different ways. First, the future consequences of current pending decisions and investments can be elaborated with respect to their primary, secondary, and unanticipated consequences. These elaborations can extend as far forward into time as is credible in terms of the local context.

Second, particular forecasting techniques can be applied to specific types of problems or forms of trends analysis. The Center for the Study of Social Poicy at SRI has identified for the U.S. Army Engineer institute for Water Resources 31 commonly used forecasting methods (Mitchell, 1977). These range from trend extrapolation to Delphi polling to input/output analysis and the complex Box-Jenkins method. These techniques can be used to generate single-point estimates in stable environments or to generate a range of plausible estimated futures in more turbulent conditions.

A third approach involves the establishment of a comprehensive planning approach, including the identification of alternative scenarios for urban development that are related, in turn, to some kind of strategic policy process. Such a model is outlined in Figure 19.2.

This outline is derived in part from an "interactive normative planning" paradigm developed by Ozbekhan (1977) and others. This paradigm has a number of phases and elements.

Phase I involves the reference projections, including an assessment of current and emerging trends and realities. The problem here is the identification and analysis of the emerging trends that might substantially affect some of the more traditional and dominant trends. Commuter transit behavior might be an example of an area where emerging behavior might differ substantially from historical transportation activities. Another significant element must be the

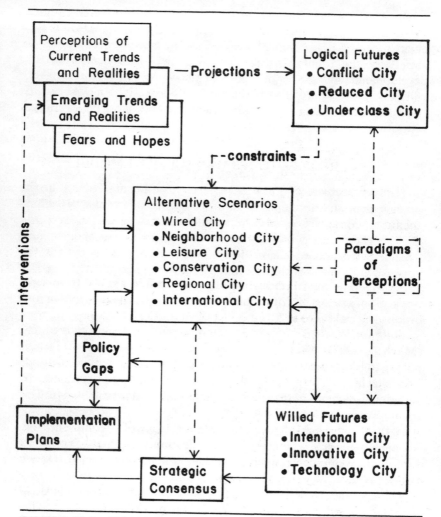

FIGURE 19.2   Urban Futures Management Model

assessment of the hopes, fears, fantasies, and values associated with urban life in particular cities.

From this kind of reference analysis one can project a number of logical futures that can be described as a dynamic projection of the status quo. Much traditional urban planning ended at this point.

A futures-driven planning model goes on to develop in Phase II a set of normative pictures of the future described as "willed futures." These can be developed as a series of plausible scenarios, documented by an analysis of emerging trends and an assessment of technological opportunities, and subject to various contraints (Shostak, 1981).

Another approach to developing a set of willed futures is to use the perceptions of a representative group of community leaders in a small workshop setting employing the Myers-Briggs Type Indicator instrument (Gappert, 1982).

The Myers-Briggs approach, which is organized around the systematic identification of different cognitive styles of perception and decision making, helps to establish different paradigms for urban change and development. The four most commonly developed paradigms usually represent (1) a strategic reconstruction orientation, (2) a pragmatic humanist viewpoint, (3) a technocratic perspective, and (4) a quasi-utopian solution. (Gappert, 1975).

The development of a useful set of willed futures usually depends upon the quality of the participation involved in their articulation. If the participation is meaningful, and involves the appropriate mix of people, then a Phase III can be initiated with the preparation of a "strategic consensus" by which the elements of an urban development plan can be articulated and subsequently guided.

Phase III must include the elaboration of the policy gaps between "what is" and "what is desired to be" the future of the city and its region.

A fourth phase is to organize an organizational and implementation plan that assigns specific issues or goals to particular agencies and organizations. This phase may be the most difficult to complete because it usually has to include (1) resource reallocation, (2) organizational restructuring, (3) the possible design of new organizations, and (4) perhaps new coalitions or working relationships.

Neither the open urban political decision-making process or the relatively private economic decision-making processes are particularly well suited to deal with these strategic organizational imperatives. The cities that need to seek a new strategic direction also need to acquire or develop an entirely new set of institutional competencies. This, however, is not an impossible task. It is likely that both the values and the skills necessary to achieve the attainment of one of the proposed willed futures already exist among the functional elites of the city.

A more significant problem for many cities is that the members of the ceremonial elite may not recognize the need for a strategic reformulation of urban purpose and form appropriate to a new paradigm of an intentional city, part of a new advanced industrial society, linked to a global economy. The existing leaders of a regional community need to articulate a coherent new vision. Out of this articulation (which may require new leaders) must come a popular consensus about the direction of urban change and development. As Shefrin (1980) has said, "Given the heavy influence of the technofuturist perspective, it becomes critical to reemphasize the role of people above imperatives.''

He indicates: ''An appropriate, popular, and dynamic image of the future shapes behavior, attitudes, and institutions, and thereby summons forth its own realization'' (p. 167).

Dennis Gabor (1969) also has stated the case well:

> The future will be made less by what is ''objectively true'' than by what people take to be true, how they relate that to their goals, what they try to do about it, what they are able to do about it, and what difference these efforts make for the kind of society that they will thus create [1969: 239].

The Intentional City of the future, guided by a strategic consensus and supported by an active citizenry, will be rich with innovation and replete with institutional vitality. The challenge of its achievement lies before us.

REFERENCES

GABOR, D. (1969) Inventing the Future. New York: Knopf.

GAPPERT, G. (1982) ''An inventive futures workshop employing the Myers-Briggs Type Indicator.'' Occasional Paper, Institute for Futures Studies and Research, The University of Akron.

———(1980) ''The future of urban management,'' in L. Rutter (ed.) The Essential Community. International City Management Association.

———(1975) ''Alternative agendas for urban policy and research in the post-affluent future,'' in G. Gappert and H. M. Rose (eds.) The Social Economy of Cities. Beverly Hills, CA: Sage.

GLUCK, F. et al. (1982) ''The four phases of strategic management.'' Journal of Business Strategy (Winter).

KNIGHT, R. V. (1979) ''City development in an industrial region: Detroit, a case study.'' Occasional Papers of Henry Ford Community College, Detroit (June).

MITCHELL, A. et al. (1977) Handbook of Forecasting Techniques, Part II. National Technical Information Service, U.S. Department of Commerce, Springfield, VA.

OZBEKHAN, H. (1977) ''The future of Paris: a systeems study in strategic urban planning. Philosophical Transaction of the Royal Society, London.

RUTTER, L. (1980) The Essential Community: Local Government in the Year 2000. Washington, DC: International City Management Association.

SHEFRIN, B. M. (1980) The Future of U.S. Politics in An Age of Economic Limits. Boulder, CO: Westview.

SHOSTAK, A. (1981) ''Long-range future: seven scenarios for Philadelphia's next 25 years. Philadelphia: Center for Philadelphia Studies.

About the Contributors

JAMES ALM is Assistant Professor of Economics at Syracuse University. He received his Ph.D. from the University of Wisconsin—Madison in 1980. He has published in the areas of fiscal federalism and housing finance.

DEBORAH BICKFORD is an economist and faculty fellow at the U.S. General Accounting Office, on leave from the Department of Urban Studies at the University of Akron. Her research interests include urban economic development, public finance, and federalism.

JOHN P. BLAIR, Professor and Chair, Department of Economics, Wright State University. He has written in the areas of urban futures and public economics. His recent publications include "The changing economics of the Urban promise," "Leveraging of public funds in economic Development," and "Strategies for housing the urban poor." He is currently working on a critique of the Urban Enterprise Zone proposal.

WARNER BLOOMBERG is Professor of Sociology, San Diego State University. His research interests include urban social systems and minority groups. Professor Bloomberg was coeditor of *Power, Poverty, and Urban Policy* and *The Quality of Urban Life,* earlier editions of the Urban Affairs Annual.

KATHLEEN BUTLER is a planner with the planning firm of Burgwyn and Company in Denver, Colorado. Ms. Butler has specialized in local government menagement and policy analysis throughout the Rocky Mountain West. She received her M.A. in Public Administration from the University of Colorado.

JESSE BURKHEAD is Maxwell Professor of Economics and Public Administration at Syracuse University. He is a Wisconsin Ph.D. and has published widely in a number of areas of applied public expenditure, including government budgeting, educational finance, state-local finance, and intergovernmental relations.

BEN CHINITZ is Dean, College of Management Science, University of Lowell, Lowell, MA. and he is the author of the sections on urban economics and regional economics in the McGraw-Hill Encyclopedia of Economics.

YONG HYO CHO is professor of Urban Studies and Political Science, University of Akron. His research, publications, and teaching are mainly centered on urban policy and management issues from local and intergovernmental perspectives.

LYNN CLOUGH is Research Associate at the Institute for Futures Studies and Research at the University of Akron. She received a graduate degree in Sociology from the University of Akron and was involved in the Akron Neighborhood Survey Project there. Her current research at the Institute has focused on strategic planning for human services in urban areas.

JOSEPH F. COATES is president of J. F. Coates, Inc., a policy research organization serving the public and private sectors in futures research, public policy, and technological decision making. He holds adjunct appointments at George Washington University and American University in Washington, D.C.

FRANK COSTA is Director of the Center for Urban Studies at the University of Akron. He is also Associate Professor of Urban Studies. He received his Ph.D. from the University of Wisconsn in 1974.

RICHARD FLEISCHMAN, FAIA, is the senior partner in Richard Fleischman Architects, Inc., a fellow in the American Institute of Architecture, and a member of the American Planning Association. He is extremely cognizant of the responsibility of designing not only a building as architecture but architecture for the urban environment. As a consultant to the City of Cleveland City Planning Commission, he has inspired and guided the development of the urban design guidelines for downtown Cleveland.

GARY GAPPERT is director of the Institute for Futures Studies and Research at the University of Akron. Holding a Ph.D degree in Economics from Syracuse University, he is the author of *Post-Affluent America* and editor of a previous urban affairs annual, *The Social Economy of Cities*. He has served as the assistant commissioner of education in New Jersey and as director of urban development for Research for Better Schools, Inc.

LARRY HIRSCHHORN is a lecturer in the social systems science program at the Wharton School, University of Pennsylvania. He is also on the staff of the affiliated Management and Behavioral Sciences Center. His research interests are in the areas of futurism, quality of work life and the political economy of decline and retrenchment. His book, *Beyond Mechanization: Flexibility and the Theory of Post-Industrial Technology*, will be published by MIT Press.

RICHARD V. KNIGHT is an economist specializing in the development of city-regions. His particular interest, stemming from his work in the Greater Cleveland Region, is the development of world industrial cities. He had a background in civil engineering, market research, and economic research in industry before earning a Ph.D. in Economics from the University of London, 1972.

Findings from an extensive research program begun at the Conservation of Human Resources Project at Columbia University and continued at Case Western Reserve University and Cleveland State University have been published in six books, including: *The Metropolitan Economy*, (1970); *Employment Expansion and Metropolitan Trade*, (1973) and *Suburbanization and the City*, (1976). Currently, he is a visiting professor at the School of Urban and Regional Studies, University of New Orleans, and is working with the National Research Council's Committee on National Urban Policy on the implications for urban policy of structural changes in the economy. Dr. Knight is also affiliated with the Institute for Futures Studies and Research at the University of Akron.

RODRIGO MARTÍNEZ-SANDOVAL is Associate Professor of Political Science and Sociology at Universidad Autonoma de Baja California (Mexicali). He is on a one-year leave of absence to study border problems at Centro de Estudios Fronterizos del Norte de Mexico (Tijuana).

SALLY ENGLE MERRY is Assistant Professor of Anthropology at Wellesley College. She has written in the fields of dispute processing, urban social organizations, crime and the fear of crime, and legal pluralism. She is the author of *Urban Danger: Life in a Neighborhood of Strangers* (Temple University Press). She is currently conducting research on community mediation.

DAVID J. O'BRIEN is Associate Professor of Sociology at the University of Akron. His research interests include urban neighborhoods (*Neighborhood Organization and Interest-Group Processes*, Princeton University Press, 1975), intergenerational changes in ethnic identification and community participation, and the adaptation of sociological theory to meet the unique parameters of the American experience.

JAMES RICHARDSON is Professor of Urban Studies and History at the University of Akron. He received his Ph.D. from New York University in 1961. He is the author of *Urban Police in the United States*.

JON VAN TIL is Associate Professor and Departmental Chair in Urban Studies and Community Development at the Rutgers University Camden College. For the past four years Van Til's research has focused on the societal impact of changing patterns of energy pricing and availability. He has published several articles in this area, and his book *Living with Energy Shortall: A Future for American Cities and Towns*, will be published by Westview Press in 1982.

CHARLES VEHORN is an economist with the U.S. General Accounting Office and Professorial Lecturer in public finance at Georgetown University. His research interests include urban public finance, economic development, and health economics. He has published articles in National Tax Journal, Public Health Reports, and Research Policy.

HAROLD M. ROSE has been professor of Urban Affairs and Geography at the University of Wisconsin—Milwaukee since 1962. His research is focused on black residential mobility and black migration patterns. More recently, attention has been paid to quality of life in black residential environments. His most recent research project involved a six-city investigation of black homicide patterns.

SUSAN SAEGERT is an Associate Professor of Environmental Psychology at the City University of New York—Graduate Center, and she also served as the first director of the Center for the Study of Women in Society. Her most recent research has included a survey of the downtown Denver workforce assessing the demand for downtown housing in the light of changing lifestyles. Recent articles include "Masculine Cities and Feminine Suburbs: Polar Ideas, Contradictory Realities" in *Women and the American City*, and "Environment and Childrens' Well-Being" in the Handbook of Medical Psychology.

ARTHUR SHOSTAK has been an applied sociologist, urbanologist, and futurist since 1961. He has published *Modern Social Reforms, Putting Sociology to Work, Sociology in Action*, eight other books, and over 80 articles. He is no researching urban future scenarios for Philadelphia, the changing nature of urban life for the nation's blue-collarites, and the social taboos that help determine the quality of life of Americans.